# MATHEMATICAL

# groups

## Tony Barnard & Hugh Neill

**TEACH YOURSELF BOOKS**

Long-renowned as the authoritative source for self-guided learning – with more than 30 million copies sold worldwide – the *Teach Yourself* series includes over 200 titles in the fields of languages, crafts, hobbies, sports, and other leisure activities.

A catalogue record for this title is available from the British Library

*Library of Congress Catalog Card Number: 96–071078*

First published in UK 1996 by Hodder Headline Plc, 338 Euston Road, London NW1 3BH

First published in US 1996 by NTC Publishing Group, 4255 West Touhy Avenue, Lincolnwood (Chicago), Illinois 60646 – 19975 U.S.A.

Printed in Great Britain by Cox & Wyman Ltd, Reading, Berkshire.

Impression number    14 13 12 11 10 9 8 7 6 5 4 3 2 1
                         2000 1999 1998 1997 1996

# *Contents*

# *Preface*

This book covers the usual material which is found in a first course on groups. The first three chapters are preliminary. Chapter 4 establishes a number of results about integers which will be used freely in the remainder of the book. The book gives a number of examples of groups and subgroups, including permutation groups, dihedral groups and groups of residue classes. The book goes on to study cosets and finishes with the first isomorphism theorem.

Very little is assumed as background knowledge on the part of the reader. Some facility in algebraic manipulation is required, and a working knowledge of some of the properties of integers, such as knowing how to factorise integers into prime factors.

The book is intended for those who are working on their own, or with limited access to other kinds of help. The authors have therefore included a number of features which are designed to help the reader who is working without other support.

Throughout the book, there are 'asides' written in *shaded italics*, which are designed to help the reader by giving an overview, or by clarifying detail. For example, sometimes the reader is told where a piece of work will be used and if it can be skipped until later in the book, and sometimes a connection is made which otherwise might interrupt the flow of the text.

The book includes very full proofs and complete answers to all the questions. Moreover, the proofs are laid out so that at each stage the reader is made aware of the purpose of that part of the proof. This approach to proofs is in line with one of our aims which is to help students with the transition from concrete to abstract mathematical thinking. Much of the student's previous work in mathematics is likely to have been computational in character: differentiate this, solve that, integrate the other, with very little deductive work being involved. But pure mathematics is about proving things, and care has been taken to give the student as much support as possible in learning how to prove things.

New terminology is written in bold type whenever it appears.

At the end of each chapter a set of key points contained in the chapter are summarised in a section entitled 'What you should know'. These sections are included to help readers to recognise the significant features for revision purposes.

Tony Barnard
Hugh Neill

April, 1996

# **1**

# *Proof*

## 1.1 THE NEED FOR PROOF

Proof is the essence of mathematics. It is a subject in which you build secure foundations, and from these foundations, by reasoning, deduction and proof, you deduce other facts and results which you then know are true, not just for a few special cases, but always.

For example, suppose that you notice that when you multiply three consecutive whole numbers such as $1 \times 2 \times 3 = 6$, $2 \times 3 \times 4 = 24$ and $20 \times 21 \times 22 = 9240$, the result is always a multiple of 6. You may make a conjecture that the product of three consecutive whole numbers is always a multiple of 6, and you can check it for a large number of cases. But you cannot assert correctly that the product of three consecutive whole numbers is always a multiple of 6 until you have provided a convincing argument that it is true no matter which three consecutive numbers you take.

For this example, a proof may consist of noting that if you have three consecutive numbers, one (at least) must be a multiple of 2 and one must be a multiple of 3, so the product is always a multiple of 6. This statement is now proved true whatever whole number you start with.

Arguing from particular cases does not constitute a proof. The only way that you can prove a statement by arguing from particular cases is by ensuring that you have examined every possible case. Clearly, when there are infinitely many possibilities, this cannot be done by examining each one in turn.

Similarly, young children will 'prove' that the angles of a triangle add up to 180° by cutting the corners off a triangle and showing that if they are placed together as in Fig. 1.1 they make a straight line, or they might measure the angles of a triangle and add them up. But, even allowing for inaccuracies of measuring, neither of these methods constitutes a proof; by their very nature they cannot show that the angle sum of a triangle is 180° for all possible triangles.

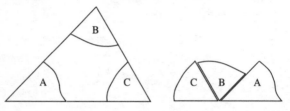

**Fig. 1.1** *'Proof' that the angles of a triangle add to 180°*

So a proof must demonstrate that a statement is true in all cases. The onus is on the prover to demonstrate that the statement is true. The argument that: 'I cannot find any examples for which it doesn't work, therefore it must be true' simply isn't good enough.

Here are two examples of statements and proofs.

## ■ *Example 1.1.1*

Prove that the sum of two consecutive whole numbers is odd.

**Proof**: Suppose that $n$ is the smaller whole number. Then $(n+1)$ is the larger number, and their sum is $n+(n+1)=2n+1$. Since this is one more than a multiple of 2, it is odd. ■

*The symbol* ■ *is there to show that the proof is complete. Sometimes, in the absence of such a symbol, it may not be clear where a proof finishes and subsequent text takes over.*

## ■ *Example 1.1.2*

Prove that if $a$ and $b$ are even, then $a + b$ is even.

**Proof**: If $a$ is even, then it can be written in the form $a = 2m$, where $m$ is a whole number. Similarly $b = 2n$ where $n$ is a whole number. Then $a + b = 2m + 2n = 2(m + n)$. Since $m$ and $n$ are whole numbers, so is $m + n$; therefore $2(m + n)$ is an even number. ■

Notice in Example 1.1.2 that the statement says nothing about the result $a + b$ when $a$ and $b$ are not both even. It simply makes no comment on any of the three cases: (1) $a$ is even and $b$ is odd; (2) $a$ is odd and $b$ is even; (3) $a$ and $b$ are both odd.

> *In fact, $a + b$ **is** even in case (3) but the statement of Example 1.1.2 says nothing about case (3).*

The same is true of general statements made in everyday life. Suppose that the statement: 'If it is raining then I shall wear my raincoat' is true. This statement says nothing about what I wear if is not raining. I might wear my raincoat, especially if it is cold or it looks like rain, or I might not.

This shows an important point about statements and proof. If you are proving the truth of a statement such as 'If $P$ then $Q$', where $P$ and $Q$ are statements such as '$a$ and $b$ are even' and '$a + b$ is even', you cannot deduce anything at all about the truth or falsity of $Q$ if the statement $P$ is not true.

## 1.2 PROVING BY CONTRADICTION

Sometimes it can be difficult to see how to proceed with a direct proof of a statement and an indirect approach may be better. Here is an example.

## ■ *Example 1.2.1*

Prove that if $a^2$ is an even number, then $a$ is an even number.

**Proof**: Suppose that $a$ is an odd number. Then $a$ can be written in the form $a = 2n + 1$, where $n$ is a whole number. Then $a^2 = (2n + 1)^2$, that is $a^2 = 4n^2 + 4n + 1 = 2(2n^2 + 2n) + 1$, so $a^2$ is 1 more than a multiple of 2, and therefore odd. But you are given that $a^2$ is an even number, so you have arrived at a contradiction. Therefore the supposition that $a$ is an odd number is untenable. Therefore $a$ is an even number. ∎

> This is an example of **proof by contradiction**, sometimes called 'reductio ad absurdum'.

Here are two more examples of proof by contradiction.

## ■ *Example 1.2.2*

Prove that $\sqrt{2}$ is irrational.

> The statement means that $\sqrt{2}$ cannot be written in the form $a/b$ where $a$ and $b$ are whole numbers.

**Proof**: Suppose that $\sqrt{2}$ is rational, that is, $\sqrt{2} = a/b$ where $a$ and $b$ are positive whole numbers with no common factors. Then, by squaring, $a^2 = 2b^2$. Since $b$ is a whole number, so is $b^2$ and $2b^2$ is an even number. Therefore $a^2$ is even, and by the result of Example 1.2.1, $a$ is even, and can therefore be written in the form $a = 2c$. The relation $a^2 = 2b^2$ can now be written as $(2c)^2 = 2b^2$ which gives $2c^2 = b^2$, showing that $b^2$ is even. Using Example 1.2.1 again, $b$ is even. You have now shown that the assumption that $\sqrt{2} = a/b$ leads to $a$ and $b$ are both even, so they both have a factor of 2. But this contradicts the assumption that $a$ and $b$ have no common factors. This is a contradiction, so the original assumption is false. Therefore $\sqrt{2}$ is irrational. ∎

## ■ *Example 1.2.3*

Prove that there is no greatest prime number.

**Proof**: Suppose that there is a greatest prime number $p$. Consider the number $m = (1 \times 2 \times 3 \times 4 \times \ldots \times p) + 1$. From its construction, $m$ is not divisible by 2, or by 3, or by 4, or by any whole number up to $p$, all these numbers leaving a remainder of 1 when divided into $m$. But, every number has prime factors. It follows that $m$ must be divisible by a prime greater than $p$, contrary to hypothesis. ∎

## 1.3 IF, AND ONLY IF

Sometimes you will be asked to show that a statement $P$ is true, if, and only if, another statement $Q$ is true. For example, prove that the product of two numbers $m$ and $n$ is even if, and only if, at least one of $m$ and $n$ is even.

The statement '$P$ is true, if, and only if, $Q$ is true' is a shorthand for two different statements:

if $P$ is true then $Q$ is true (i.e., $P$ is true *only if* $Q$ is true)

and

if $Q$ is true then $P$ is true (i.e., $P$ is true *if* $Q$ is true)

There are thus two different things to prove. Here is an example.

### ■ *Example 1.3.1*

Prove that the product $mn$ of two numbers $m$ and $n$ is even if, and only if, at least one of $m$ and $n$ is even.

**Proof**: *If.* Suppose that at least one of $m$ and $n$ is even. Suppose it is $m$. Then $m = 2p$ for a whole number $p$. Then $mn = 2pn = 2(pn)$, so $mn$ is even.

> *In proofs which involve 'if, and only if' the proof will generally be laid out in this way with the 'if' part first, followed by the 'only if' part.*
>
> *Here is a contradiction proof of the second result, that if $mn$ is even, then at least one of $m$ and $n$ is even.*

*Only if.* Suppose that the statement 'at least one of $m$ and $n$ is even' is false. Then $m$ and $n$ are both odd. The product of two odd numbers is odd. (You are asked to prove this statement in Exercise 1, question 1.) This is a contradiction, as you are given that $mn$ is even. Hence at least one of $m$ and $n$ is even. ■

Another way of saying the statement '$P$ is true, if, and only if, $Q$ is true', is to say that the two statements $P$ and $Q$ are **equivalent**.

> *Thus, to prove that P and Q are equivalent you have to prove that each statement can be proved from the other.*

Another way of describing equivalent statements $P$ and $Q$ is to say that $P$ is a **necessary and sufficient condition** for $Q$. For example, a necessary and sufficient condition for a number $N$ to be divisible by 3 is that the sum of the digits of $N$ is divisible by 3.

The statement '$P$ is a sufficient condition for $Q$' means

> if $P$ is true then $Q$ is true.

> *If P is true, this is enough for Q to be true.*

and the statement '$P$ is a necessary condition for $Q$' means

> if $Q$ is true then $P$ is true.

> *Q cannot be true without P also being true.*

So, once again there are two different things to prove. Here is an example.

## ■ *Example 1.3.2*

Prove that a necessary and sufficient condition for a number $N$ expressed in denary notation to be divisible by 3 is that the sum of the digits of $N$ is divisible by 3.

**Proof**: Any integer $N$ may be written in denary notation in the form $N = a_n 10^n + a_{n-1} 10^{n-1} + \ldots + a_1 10 + a_0$ where $0 \le a_i < 10$ for all $i$.

*Necessary.* If 3 divides $N$, then 3 divides $a_n 10^n + \ldots + a_1 10 + a_0$. But for all $i$, $10^i$ leaves remainder 1 on division by 3. So the remainder when $N$ is divided by 3 is $a_n + \ldots + a_0$. But as 3 divides $N$ the remainder is 0, so $a_n + \ldots + a_0$ is divisible by 3, that is, the sum of the digits is divisible by 3.

*Sufficient.* Suppose now that the sum of the digits is divisible by 3, that is, $a_n + \ldots + a_0$ is divisible by 3. Then 3 also divides the sum

$$\left( a_n + \ldots + a_0 \right) + \left( \overbrace{9 \ldots 9}^{n \; 9s} a_n + \ldots + 9a_1 \right)$$

$$= a_n\left(1 + \overbrace{9\ldots9}^{n\ 9s}\right) + \ldots + a_1(1+9) + a_0$$

$$= a_n\left(10^n\right) + \ldots + a_1(10) + a_0 = N. \ \blacksquare$$

## 1.4 DEFINITIONS

It is a (somewhat unhelpful) convention that when mathematical terms are defined, the word 'if' is used when 'if, and only if,' is meant.

For example, in Chapter 2, there is a definition of equality between two sets which says that 'two sets $A$ and $B$ are equal if they have the same members'. Although this is a true statement, in order to work with the notion of equality of sets, you need the stronger statement that 'two sets $A$ and $B$ are equal if, and only if, they have the same members'.

*You will be reminded of this convention when the case arises later.*

## 1.5 PROVING THAT SOMETHING IS FALSE

You will sometimes need to show that a statement is false. For example, such a statement might be 'Prime numbers are always odd'. To show a statement is false, you need only to find one example which contradicts or runs counter to the statement. In this case, there is only one example, namely, 2. So the statement is false.

In this case, 2 is called a counterexample.

You can show that the statement 'Odd numbers are always prime' is false by producing the counterexample 9, which is odd, and not prime.

A particular case which shows a statement to be false is called a **counterexample**.

Sometimes there will be many counterexamples. For example, to show that the statement 'if $n$ is an integer, then $n^2 + n + 41$ is a prime number' is false, one counterexample is $n = 41$. But any non-zero multiple of 41 would also be a counterexample.

Sometimes a statement is not true, but a counterexample is difficult to find. For example, the statement 'there are no whole numbers $m$ and $n$ such that $m^2 - 61n^2 = 1$' is false, but the smallest counterexample is $m = 1\ 766\ 319\ 049$ and $n = 226\ 153\ 980$.

## 1.6 CONCLUSION

This chapter has been about proof, and the fact that considering special cases never constitutes a proof, unless you consider all the possible special cases. However, do not underestimate considering special cases; sometimes they can show you the way to a proof of something. But don't forget that you can never be certain that something is true if your trust in it depends only on having considered special cases.

## WHAT YOU SHOULD KNOW

■ The meaning of 'proof by contradiction'.

■ That you cannot prove something by looking at special cases, unless you can look at all the special cases.

■ How to prove results which involve 'if, and only if'.

■ How to prove that two statements are equivalent.

■ The meaning of 'a necessary and sufficient condition for'.

■ How to use counterexamples.

## EXERCISE I

**1** Prove that the product of two odd numbers is odd.

**2** Use proof by contradiction to show that it is not possible to find positive whole numbers $m$ and $n$ such that $m^2 - n^2 = 6$.

**3** Prove that a number plus its square is always even.

**4** Fig 1.2 shows four playing cards. Two are face up, and two are face down. Each card has either a chequered pattern or a striped pattern on its reverse side.

**Fig. 1.2**

Which cards would you need to turn over in order to determine whether the statement: "Each card with a striped pattern on its reverse side is a diamond", is true?

**5**   Return to the proof in Example 1.2.2 that $\sqrt{2}$ is irrational. Where does this proof break down if you replace 2 by 4?

**6**   Prove that the statement 'if $x < 1$ then $x^2 < 1$' concerning real numbers, is false.

**7**   Prove that if $\sqrt{a+b} = \sqrt{a} + \sqrt{b}$, then $a = 0$ or $b = 0$.

**8**   Prove that the product $pq$ of two integers $p$ and $q$ is odd if, and only if, $p$ and $q$ are both odd.

**9**   Prove that a necessary and sufficient condition for an integer $N$ expressed in denary notation to be divisible by 9 is that 9 divides the sum of the digits of $N$.

# 2

# *Sets*

## 2.1 WHAT IS A SET?

**Definition**: A **set** is a collection of things; the things are called **elements** or **members** of the set.

In this book, sets will be denoted by capital letters such as $A$ or $B$.

A set is determined by its members. To define a set you can either list its members or you can describe it in words, provided you do so unambiguously. When you define a set, there should never be any uncertainty about what its elements are.

For example, you could say that the set $A$ consists of the numbers 1, 2 and 3; or alternatively, that 1, 2 and 3 are the elements of $A$. Then it is clear that $A$ consists of the three elements 1, 2 and 3, and that 4 is not an element of $A$.

The symbol $\in$ is used to designate 'is a member of' or 'belongs to'; the symbol $\notin$ means 'is not a member of' or 'does not belong to'. So $1 \in A$ means that 1 is a member of $A$, and $4 \notin A$ means that 4 is not a member of $A$.

## 2.2 EXAMPLES OF SETS: NOTATION

There are some sets which will be used so frequently that it is helpful to have some special names for them.

### ■ *Example 2.2.1*

The set consisting of all the **integers**, (that is, whole numbers), $\dots, -2, -1, 0, 1, 2, \dots$ is denoted by $\mathbf{Z}$.

*Z is for Zahlen, the German for numbers.*

Using the notation introduced in the previous section you can write $-5 \in \mathbf{Z}$ and $\frac{1}{2} \notin \mathbf{Z}$.

### ■ *Example 2.2.2*

The set consisting of all the **natural numbers** (positive integers) $1, 2, \dots$ is denoted by $\mathbf{N}$.

You can write $-5 \notin \mathbf{N}$ and $5 \in \mathbf{N}$.

*It is easy to forget whether 0 does or does not belong to $\mathbf{N}$, and different books often define $\mathbf{N}$ differently in this respect. Be warned!*

### ■ *Example 2.2.3*

The **real numbers** will be called $\mathbf{R}$, and the **complex numbers** will be called $\mathbf{C}$. The notation $\mathbf{R}^+$ means the positive real numbers. The notations $\mathbf{R}^*$ and $\mathbf{C}^*$ will be used to mean the non-zero real and complex numbers respectively.

## 2.3 DESCRIBING A SET

When you list the members of a set, it is usual to put them into curly brackets $\{\ \}$, often called braces. For example, the set $A$ consisting of the elements 2, 3 and 4 can be written $A = \{2, 3, 4\}$. The order in which the elements are written doesn't matter. The set $\{2, 4, 3\}$ is identical to the set $\{2, 3, 4\}$, and $A = \{2, 3, 4\} = \{2, 4, 3\}$.

If you wrote $A = \{2,2,3,4\}$, it would be the same as saying that $A = \{2,3,4\}$. A set is determined by its distinct members, and any repetition in the list of members can be ignored.

Using braces, you can write $\mathbf{Z} = \{\dots,-2,-1,0,1,2,\dots\}$.

Another way to describe a set involves specifying properties of its members. For example,

$$A = \{n \in \mathbf{Z} : 2 \leq n \leq 4\}$$

means that $A$ is the set of all natural numbers $n$ such that $2 \leq n \leq 4$. The symbol : means 'such that'. To the left of the symbol : you are told a typical member of the set, while to the right you are given a condition which the element must satisfy.

So in $A$, a typical element is an integer; and the integer must lie between 2 and 4 inclusive. Therefore $A = \{2,3,4\}$.

## ■ *Example 2.3.1*

The set of rational numbers, $\mathbf{Q}$, (for quotients), is

$$\mathbf{Q} = \left\{ \frac{m}{n} : m,n \in \mathbf{Z}, n \neq 0 \right\}$$

$\mathbf{Q}^+$ denotes the positive rationals and $\mathbf{Q}*$ denotes the non-zero rationals.

## 2.4 SUBSETS

**Definition**: If all the members of a set $A$ are also members of another set $B$, then $A$ is called a **subset** of $B$. In this case you write $A \subseteq B$.

You can see from the definition of subset that $A \subseteq A$ for any set $A$.

*Notice that the notation $A \subseteq B$ suggests the notation for inequalities, $a \leq b$. This analogy is intentional and helpful. However, you mustn't take it too far: for any two numbers you have either $a \leq b$ or $b \leq a$, but the same is not true for sets. For example, for the sets $A = \{1\}$ and $B = \{2\}$ neither $A \subseteq B$ nor $B \subseteq A$ is true.*

Some writers use $A \subset B$ to mean that $A$ is a **proper subset** of $B$, that is $A \subseteq B$ and $A \neq B$. This notation will not be used in this book.

**Definition**: Two sets $A$ and $B$ are called **equal** if they have the same members.

> *Remember the comment at the end of Section 1.4. 'Two sets A and B are called equal if they have the same members' is an example where 'if' is used, but 'if, and only if' is meant.*

## ■ *Example 2.4.1*

If $A = \{$letters of the alphabet$\}$,

    $B = \{x \in A : x$ is a letter of the word '*stable*'$\}$,

    $C = \{x \in A : x$ is a letter of the word '*bleats*'$\}$,

    $D = \{x \in A : x$ is a letter of the word '*Beatles*'$\}$

    $E = \{x \in A : x$ is a letter of the word '*beetles*'$\}$

Then $B = C = D = \{a, b, e, l, s, t\}$. However, $D \neq E$.

In fact, $E = \{b, e, l, s, t\}$ is a proper subset of $D$.

## 2.5 VENN DIAGRAMS

Figure 2.1 shows a way of picturing sets.

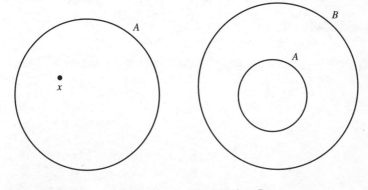

**Fig. 2.1** $x \in A$          **Fig. 2.2** $A \subseteq B$

The set $A$ is drawn as a circle (or an oval), and the element $x$, which is a member of $A$, is drawn as a point inside $A$. So Fig. 2.1 represents $x \in A$.

In Fig. 2.2, every point inside $A$ is also inside $B$, so this represents the statement that $A$ is a subset of $B$, or $A \subseteq B$.

*These diagrams can be helpful for understanding, seeing and suggesting relationships, but be warned; they can also sometimes be misleading. For example, in Fig. 2.2, the question of whether or not $A = B$ is left open. The fact that on the diagram there are points outside $A$ and inside $B$ does not mean that there are necessarily elements in $B$ which are not in $A$. Take care when using diagrams!*

## 2.6 INTERSECTION AND UNION

**Definition**: Suppose that $A$ and $B$ are any two sets. Then the **intersection** of $A$ and $B$, written $A \cap B$ and pronounced 'A intersection $B$' is the set

$$A \cap B = \{x : x \in A \text{ and } x \in B\}$$

It is clear from the definition that $A \cap B = B \cap A$.

**Definition**: The **union** of $A$ and $B$, written $A \cup B$ and pronounced 'A union $B$' is the set

$$A \cup B = \{x : x \in A \text{ or } x \in B \text{ or both}\}$$

It is also clear from the definition that $A \cup B = B \cup A$.

Figs. 2.3 and 2.4 illustrate the union and intersection of sets $A$ and $B$.

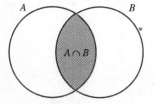

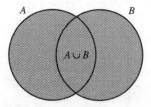

**Fig. 2.3** *The set* $A \cap B$                     **Fig. 2.4** *The set* $A \cup B$

# ■ *Example 2.6.1*

Suppose that $A = \{2,3,4\}$ and $B = \{4,5,6\}$. Then $A \cap B = \{4\}$ and $A \cup B = \{2,3,4,5,6\}$.

Notice that if $A = \{2,3,4\}$ and $B = \{5,6,7\}$, then $A \cap B$ has no members. You write this as $A \cap B = \{\ \}$, where the right-hand side is an empty pair of curly brackets. This set is called the empty set. The special symbol $\varnothing$ is used, so in this case $A \cap B = \varnothing$.

**Definition**: The set with no members is called the **empty set**; it is denoted by $\varnothing$.

Two sets $A$ and $B$ are said to be **disjoint**, if they have no members in common. Thus $A$ and $B$ are disjoint if, and only if, $A \cap B = \varnothing$.

The question of whether $\varnothing$ is a subset of $A$ or $B$ is somewhat awkward. For instance, is it true that if $x \in \varnothing$ then $x \in A$? As there is no member $x$ which belongs to $\varnothing$ it is certainly not possible to find an $x$ to show that the statement 'if $x \in \varnothing$ then $x \in A$' is false. So, conventionally, $\varnothing$ is regarded as a subset of every set $A$.

# ■ *Example 2.6.2*

Suppose that $Q$ is the set of plane quadrilaterals and $T$ is the set of all triangles. Then $Q$ and $T$ are disjoint and $Q \cap T = \varnothing$.

# ■ *Example 2.6.3*

Suppose that $D$ is the set of all rhombuses, and $R$ is the set of rectangles. Then $D \cap R$ is the set of squares.

## 2.7 PROVING THAT TWO SETS ARE EQUAL

To prove that two sets $A$ and $B$ are equal, you often prove separately that $A \subseteq B$ and $B \subseteq A$.

*At first sight this may seem as if a simple task has been replaced with two less simple tasks, but in practice, it gives a method of proving that two sets are equal, namely by proving that every member of the first is a member of the second, and vice versa.*

Here is an example suggested by the Venn diagram in Fig. 2.5.

### ■ *Example 2.7.1*

Prove that if $A \subseteq B$ then $A \cap B = A$.

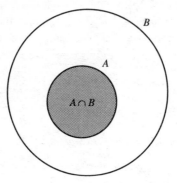

**Fig. 2.5**

**Proof:**

> In this case the hypothesis that $A \subseteq B$, tells you that if $x \in A$ then $x \in B$. This will be used in appropriate places in the proof that $A \cap B = A$.
>
> The proof that $A \cap B = A$ has two parts: first, to prove that $A \cap B \subseteq A$; and secondly, to prove that $A \subseteq A \cap B$.

*First part.* Suppose first that $x \in A \cap B$. Then $(x \in A$ and $x \in B)$. It follows that $x \in A$, so $A \cap B \subseteq A$.

> When you have a composite entity, such as $A \cap B$, sometimes you should think of it just as one compressed item, and at other times you may need to unpack it and work with the pieces.

*Second part.* Now suppose that $x \in A$. From the hypothesis if $x \in A$ then $x \in B$, so $(x \in A$ and $x \in B)$. Therefore $x \in A \cap B$. Hence $A \subseteq A \cap B$.

The two results $A \cap B \subseteq A$ and $A \subseteq A \cap B$ together imply that $A \cap B = A$. ■

It is also true that if $A \cap B = A$, then $A \subseteq B$. This is proved as Example 2.7.2.

## ■ *Example 2.7.2*

Prove that if $A \cap B = A$, then $A \subseteq B$.

**Proof**:

> It can sometimes appear hard to know where to start in a proof like this. To prove that $A \subseteq B$, you have to start with $x \in A$, and from it deduce that $x \in B$. Somewhere along the way you will use the hypothesis $A = A \cap B$.

If $x \in A$ then, from the hypothesis, $x \in A \cap B$, so $x \in A$ and $x \in B$. It follows that $x \in B$. Hence if $x \in A$ then $x \in B$, so $A \subseteq B$. ■

Notice that Examples 2.7.1 and 2.7.2 together show that the two statements $A \subseteq B$ and $A \cap B = A$ are equivalent because each can be deduced from the other.

## WHAT YOU SHOULD KNOW

- ■ That a set needs to be well-defined in the sense that you can tell clearly whether or not a given element is a member of the set in question.

- ■ How to list the members of a set.

- ■ The meanings of the symbols $\in$, $\notin$, : and the bracket notation for sets.

- ■ The meaning of and notation for subsets.

- ■ The meaning of and notation for the empty set.

- ■ The meaning of and notation for intersection and union of two sets.

- ■ How to prove that two sets are equal.

## Exercise 2

**1** Which of the following sets is well defined?

(1) The set of prime numbers.

(2) People in the world whose birthday is on 1 April.

(3) $A = \{x \in \mathbf{Z} : x \text{ is the digit in the 1000th}$

place in the decimal expansion of $\pi\}$

**2** Mark each of the following statements true or false.

(1) $0 \in \mathbf{Q}$

(2) $0 \in \mathbf{Z}^*$

(3) A set may have just one element.

**3** List the members of the set $P = \{x \in \mathbf{Z} : |x| < 3\}$. Does $-3 \in P$ ?

**4** Explain why $\mathbf{Z} \subseteq \mathbf{Q}$.

**5** The notation $2\mathbf{Z}$ is used to mean the set $\{2x : x \in \mathbf{Z}\}$. Decide whether $2\mathbf{Z} \subseteq \mathbf{Z}$ or $\mathbf{Z} \subseteq 2\mathbf{Z}$ or neither or both is true.

In each of the following examples, $A$, $B$ and $C$ are any sets.

**6** Prove that the statements $A \cup B = B$ and $A \cap B = A$ are equivalent.

**7** Prove that $A \cap (B \cup C) = (A \cap B) \cup (A \cap C)$.

**8** Mark each of the following statements true or false for general sets $A$ and $B$.

(1) $\varnothing \subseteq A$

(2) $A \subseteq (A \cap B)$

(3) $A \subseteq (A \cup B)$

(4) $A \subseteq A$

(5) $A \in A$

(6) $A$ is a proper subset of $A$.

(7) $(A \cup B) \subseteq (A \cap B)$

**9** Let $A = \{x \in \mathbf{R} : |x| < 3\}$ and $B = \{x \in \mathbf{R} : |x - 1| < 2\}$. Prove that $B \subseteq A$.

**10** Let $A$ be a set with $n$ elements. Prove that $A$ has $2^n$ subsets.

# 3

## Binary operations

### 3.1 INTRODUCTION

In the treatment of sets so far in this book, you have not been concerned with any relationships between the elements of the sets.

However, there may be relationships between elements: for example, in the set $\mathbf{Z}$ you know that you can multiply elements (which are just numbers) together, or you can add them, and the result is still an element of $\mathbf{Z}$.

In $\mathbf{R}$, the difference between two numbers $x$ and $y$, defined by $|x - y|$, is an example of combining two numbers in a set and producing another number in the set, though in this case the new numbers produced are never negative.

This chapter is about such rules and some of their properties.

### 3.2 BINARY OPERATIONS

Here are some other examples of rules.

For example, in $2 + 3 = 5$, two elements 2 and 3 taken from $\mathbf{Z}$ have been combined using the operation +; the result, 5, is a member of $\mathbf{Z}$.

Suppose now that you are working with the elements of **Z** and the rule of division, ÷. This time $6 \div 2 = 3$ is an element of the set **Z**, but $5 \div 2$ has no meaning in **Z**, because there is no integer $n$ for which $5 = 2n$. In this case, the rule ÷ sometimes gives you a member of the set and sometimes does not.

In **R**\*, where division $x \div y$ is defined for all members $x$ and $y$ of **R**\*, the order of the elements matters. In general $x \div y$ is not equal to $y \div x$.

**Definition**: A **binary operation** ∘ on a set $A$ is a rule which assigns to each ordered pair of elements in $A$ exactly one element of $A$.

*Notice that it is hard to say precisely what a 'rule' is. In this case, a rule is to be understood as no more than an association of a single element of A with an ordered pair of elements of A.*

So multiplication and addition in **Z** are binary operations, and so is difference in **R** and division in **R**\*. However, the rule ÷ on **Z** is not a binary operation, because division is not defined for every pair of elements in **Z**.

Sometimes it is convenient to use the word 'operation' instead of binary operation.

## 3.3 EXAMPLES OF BINARY OPERATIONS

### ■ *Example 3.3.1*

In **N**, suppose that $a \circ b$ is the highest common factor of $a$ and $b$. Then $4 \circ 14 = 2$, and $3 \circ 7 = 1$. In this case, ∘ is a binary operation on **N**.

### ■ *Example 3.3.2*

In **Z**, suppose that $a \circ b$ is the result of subtracting $b$ from $a$, that is $a - b$. Then the result is always an integer, and ∘ is a binary operation on **Z**.

Notice that in Example 3.3.1, the order of the elements $a$ and $b$ does not matter. The highest common factor of $a$ and $b$ is the same as the highest common factor of $b$ and $a$ for all $a,\ b \in \mathbf{Z}$, so $a \circ b = b \circ a$ for all $a,\ b \in \mathbf{Z}$.

In Example 3.3.2 however, the order of $a$ and $b$ does matter. For example, it is not true that $2 - 3 = 3 - 2$.

**Definition**: A binary operation $\circ$ on a set $S$ is **commutative** if $a \circ b = b \circ a$ for all $a,\ b \in S$.

Suppose that you have an expression of the form $a \circ b \circ c$ where $\circ$ is a binary operation on a set $S$, and $a$, $b$ and $c \in S$. Then you can evaluate $a \circ b \circ c$ in two different ways, either as $(a \circ b) \circ c$ or as $a \circ (b \circ c)$.

Sometimes these two ways of calculating $a \circ b \circ c$ give the same result. For example, using the binary operation $+$ on $\mathbf{Z}$ allows you to say that $(2 + 3) + 4 = 2 + (3 + 4)$. In general $a + (b + c) = (a + b) + c$ for all $a$, $b$ and $c \in \mathbf{Z}$. In this case, it doesn't matter in which order you calculate the result. It follows that you can write $a + b + c$ without ambiguity.

However, sometimes it does matter how you work out $a \circ b \circ c$. For example, using the binary operation $-$ on $\mathbf{Z}$, $(2 - 3) - 4 = -5$ and $2 - (3 - 4) = 3$, so it does matter which way you bracket the expression $2 - 3 - 4$.

**Definition**: A binary operation $\circ$ on a set $S$ is **associative** if $(a \circ b) \circ c = a \circ (b \circ c)$ for all $a$, $b$ and $c \in S$.

For a binary operation which is associative, it is usual to leave out the brackets and to write simply $a \circ b \circ c$. You can then use either method of bracketing to calculate it.

## 3.4 TABLES

It is sometimes useful to be able to specify a binary operation by using a table.

Consider the set of numbers $X = \{2, 4, 6, 8\}$. Define a binary operation $\circ$ between the elements of $X$ by $a \circ b =$ the remainder after dividing $a \times b$ by 10.

Fig 3.1 shows a table with the results of this binary operation.

|   |   | Second number | | |
|---|---|---|---|---|
| ∘ | 2 | 4 | 6 | 8 |
| 2 | 4 | 8 | 2 | 6 |
| 4 | 8 | 6 | 4 | 2 |
| 6 | 2 | 4 | 6 | 8 |
| 8 | 6 | 2 | 8 | 4 |

First number (labels for rows)

**Fig. 3.1**

The shaded cell in Fig. 3.1 shows that $4 \circ 8 = 2$.

In this book, the following notation for tables will be used.

(The element in the $i$th row) ∘ (the element in the $j$th column)
        = (the element in the $i$th row and $j$th column).

The table in Fig. 3.2 shows how this notation works.

| ∘ | $a$ | $b$ | $c$ |
|---|---|---|---|
| $a$ | $a \circ a$ | $a \circ b$ | $a \circ c$ |
| $b$ | $b \circ a$ | $b \circ b$ | $b \circ c$ |
| $c$ | $c \circ a$ | $c \circ b$ | $c \circ c$ |

**Fig. 3.2**

The table in Fig. 3.3 shows how a table can define a binary operation: this particular example defines a binary operation on the set $S = \{a, b, c\}$.

| ∘ | $a$ | $b$ | $c$ |
|---|---|---|---|
| $a$ | $b$ | $c$ | $a$ |
| $b$ | $b$ | $a$ | $c$ |
| $c$ | $c$ | $b$ | $c$ |

**Fig. 3.3**

The fact that every cell in the table is filled by exactly one element of $S$ tells you that $\circ$ is a binary operation on $S$.

You can see from the table that $a \circ b = c$ and that $b \circ a = b$; the binary operation $\circ$ is therefore not commutative.

Tables are useful for showing how binary operations work when the sets have only a few elements; they can get tedious when there are many elements in the set.

## 3.5 TESTING FOR BINARY OPERATIONS

You need to take care when you define a binary operation and when you test whether a given rule is actually a binary operation on a set $S$.

You need to be sure that:

● there is at least one element assigned for each pair of elements in the set $S$
● there is at most one element for each pair of elements in $S$
● the assigned element is actually in $S$.

When these three properties hold, the binary operation is said to be **well defined**.

### ■ *Example 3.5.1*

Define $\circ$ on $\mathbf{R}$ by $a \circ b = a/b$. Then $\circ$ is not a binary operation because no element is assigned to the pair $a = 1$, $b = 0$.

Define $\circ$ on $\mathbf{Q}^*$ by $a \circ b = a/b$. Then $\circ$ is properly defined and $a \circ b$ is in $\mathbf{Q}^*$ for every pair $(a,b) \in \mathbf{Q}^*$; so $\circ$ is a binary operation on $\mathbf{Q}^*$.

Define $\circ$ on $\mathbf{Q}^+$ by $a \circ b = \sqrt{ab}$, where the positive square root is taken. Then no element is assigned to the pair $a = 1$ and $b = 2$; so $\circ$ is not a binary operation on $\mathbf{Q}^+$.

Define $\circ$ on $\mathbf{Z}$ by $a \circ b$ is equal to the least member of $\mathbf{Z}$ which is greater than both $a$ and $b$; then $\circ$ is a binary operation on $\mathbf{Z}$.

Define $\circ$ on $\mathbf{R}$ by $a \circ b$ is equal to the least member of $\mathbf{R}$ greater than both $a$ and $b$. Then $\circ$ is not defined for the pair $a = 0$ and $b = 0$ in $\mathbf{R}$

because there is no least positive real number. So ∘ is not a binary operation on **R**.

There is another piece of language which you might come across associated with binary operations.

A binary operation ∘ on a set $S$ is said to be **closed** if the element assigned to $a \circ b$ is in $S$ for all pairs of elements $a$ and $b$ in $S$.

A binary operation ∘ on a set $S$ is automatically closed because part of the definition of a binary operation is that the element assigned to $a \circ b$ is in $S$ for each pair $(a,b) \in S$.

## WHAT YOU SHOULD KNOW

- What a binary operation is.

- How to test whether a given operation is a binary operation or not.

- The meaning of the term 'closed'.

## EXERCISE 3

**1**  Which of the given operations are binary operations on the given set? For each operation which is not a binary operation, give one reason why it is not a binary operation.

(1)  $(\mathbf{Z}, \times)$

(2)  $(\mathbf{N}, \Diamond)$, where $a \Diamond b = a^b$

(3)  $(\mathbf{R}, \div)$

(4)  $(\mathbf{Z}, \Diamond)$

(5)  $(\mathbf{Z}, \circ)$, where $a \circ b = a$ for all $a,\ b \in \mathbf{Z}$

(6)  $(\{1,3,7,9\}, \circ)$ where $a \circ b$ is the remainder when $a \times b$ is divided by 10

(7)  $(\mathbf{R}, \circ)$, where $a \circ b = 0$ for all $a,\ b \in \mathbf{R}$

(8)  $(\mathbf{C}, \circ)$, where $a \circ b = |a - b|$

(9)  $(\mathbf{M}, \times)$, where **M** is the set of matrices, and $\times$ is multiplication of matrices

(10)  $(\mathbf{M}, \circ)$, where **M** is the set of $2 \times 2$ matrices and $A \circ B$ is given by $A \circ B = \det(A - B)$

(11)  $(\mathbf{R}, \Diamond)$, where $a \Diamond b = a^b$

# 4

# *The integers*

## 4.1 INTRODUCTION

This chapter is in the nature of a diversion from the main theme of this book. But, as you will see in due course, some sets of integers form groups, and you will need to know some of the properties of the integers in order to prove certain results about groups later on.

As you already know a great deal about the integers, the approach taken is not to start from definitions of integers, but to assume most of the major properties that you already know and to prove, in logical sequence, the results which are needed. You will see that not all the proofs are given, not because they are too advanced or too difficult, but because this is really a book about groups. But it is useful to have the theorems about integers clearly stated so that they can be used later in this book.

The first property is probably obvious, but you may not have seen it stated explicitly before. It is called the **well-ordering principle**.

Suppose that $A$ is a non-empty set of positive integers. Then $A$ has a least member.

*This property is taken as an axiom. Although it looks obvious, you cannot prove it from the usual properties of integers, unless of course, you assume something equivalent to it.*

The second property is called the **division algorithm**. You know it well, but you may not have seen it expressed in this way.

Let $a$ be any integer, and let $b$ be a positive integer. Then $a$ can be written in the form $a = qb + r$ where $0 \le r < b$, and $q$ and $r$ are unique.

*The idea here is that you take the largest multiple of $b$ which is less than $a$ (call it $qb$) and let $r$ be the remainder. Then $r = a - qb$.*

## 4.2 RELATIVELY PRIME PAIRS OF NUMBERS

**Definition**: Let $a$, $b \in \mathbf{N}$. Then $a$ and $b$ are **relatively prime** if they have no common positive divisor other than 1.

**Theorem 1**: Let $a$, $b \in \mathbf{N}$. Then $a$ and $b$ are relatively prime if, and only if, there exist integers $x$, $y$ such that $ax + by = 1$.

**Proof**: *If.* Suppose that there exist integers $x$, $y$ such that $ax + by = 1$. Let $d$ be any positive integer which divides $a$ and $b$. Then $d$ divides $ax + by$, so $d$ divides 1. Hence $d = 1$.

*Only if.* Let $h$ be the least positive integer in the set $S = \{ax + by : x, y \in \mathbf{Z}\}$. Use the division algorithm to write $a = qh + r$ where $0 \le r < h$. Now $r \in S$, because $h$ can be written as $h = ax_0 + by_0$ (since $h \in S$), so $r = a - qh = (1 - qx_0)a + (-qy_0)b$. But $h$ was the least positive integer in $S$, and $0 \le r < h$. Hence $r = 0$. Therefore $a = qh$ so $h$ divides $a$. Similarly $h$ divides $b$. But the only divisor of $a$ and $b$ is 1. So $h = 1$. ∎

Following on from this theorem, you can prove the following two results.

**Theorem 2**: Let $m$ and $n$ be relatively prime. If $m$ divides $na$, then $m$ divides $a$.

**Proof**: By Theorem 1, there exist integers $x$, $y$ such that $mx + ny = 1$. Therefore $mxa + nya = a$. Since $m$ divides the left-hand side, $m$ divides the right-hand side. Therefore $m$ divides $a$. ∎

**Theorem 3**: Let $m$ and $n$ be relatively prime. If $m$ divides $k$ and $n$ divides $k$, then $mn$ divides $k$.

**Proof**: As $n$ divides $k$, you can write $k = ns$ for some $s \in \mathbf{Z}$. Therefore, by Theorem 2, since $m$ divides $ns$, $m$ divides $s$. Therefore $s = mt$ for some $t \in \mathbf{Z}$. Therefore $k = mnt$, so $mn$ divides $k$. ∎

## 4.3 PRIME NUMBERS

**Definition**: A **prime number** is an integer $p > 1$ which has no positive divisors other than 1 and $p$.

**Theorem 4**: If $p$ divides $ab$, then either $p$ divides $a$ or $p$ divides $b$.

**Proof**: Let $p$ divide $ab$, and suppose that $p$ does not divide $a$. As the only divisors of $p$ are $p$ and 1, it follows that $p$ and $a$ are relatively prime. The result then follows from Theorem 2. ∎

From these theorems it is now possible to prove the fundamental theorem of arithmetic, which you certainly already know.

**Theorem 5**: **The fundamental theorem of arithmetic.** Every positive integer can be written uniquely as a product of primes. ∎

*Theorem 4 is used in proving the uniqueness part of Theorem 5. No proof of Theorem 5 is given here.*

*Theorem 5 justifies the statement that every number has prime factors, a result that was used in the proof in Example 1.2.3 that there is no greatest prime number.*

Using Theorem 5, you can prove a result about highest common factors and least common multiples.

**Theorem 6**: For any two positive integers $a$ and $b$, if $h$ is their highest common factor and $l$ is their least common multiple, then $ab = hl$.

This is given as an exercise, and you will find a proof in the answers. ∎

## 4.4 RESIDUE CLASSES OF INTEGERS

Here is a new piece of notation. Define $a \equiv b \pmod{n}$, read as '*a* is congruent to *b*, modulo *n*', to mean that *n* divides $a - b$.

Here are some examples of this notation.

### ■ *Example 4.4.1*

$3 \equiv 24 \pmod 7$ because $3 - 24 = (-3) \times 7$.

$11 \equiv -31 \pmod 7$ because $11 - (-31) = 6 \times 7$.

$25 \not\equiv 12 \pmod 7$ because $25 - 12 = 13$, and 13 is not a multiple of 7.

**Theorem 7**: $a \equiv b \pmod n$ if, and only if, *a* and *b* leave the same remainder on division by *n*.

**Proof**: From the division algorithm you know that there are integers $q_1$ and $r_1$, such that $a = q_1 n + r_1$, where $0 \leq r_1 < n$. Similarly, there are integers $q_2$ and $r_2$ such that $b = q_2 n + r_2$, where $0 \leq r_2 < n$. On subtracting these two equations you find that

$$a - b = (q_1 n + r_1) - (q_2 n + r_2)$$
$$= (q_1 - q_2)n + (r_1 - r_2)$$

*This equation will be used in the course of the proof.*

*If.* Suppose that *a* and *b* leave the same remainder on division by *n*, that is, $r_1 = r_2$. Then $a - b = (q_1 - q_2)n$. Therefore *n* divides $a - b$, so $a \equiv b \pmod n$.

*Only if.* Suppose that $a \equiv b \pmod n$. Then there is an integer *k* such that $a - b = kn$. Therefore

$$(q_1 - q_2)n + (r_1 - r_2) = kn$$
$$(r_1 - r_2) = kn - (q_1 - q_2)n$$

But, from $0 \leq r_1 < n$ and $0 \leq r_2 < n$ you can deduce that $-n < r_1 - r_2 < n$, so $r_1 - r_2$ is a multiple of *n* lying strictly between $-n$ and *n*. The only possibility is $r_1 - r_2 = 0$ so $r_1 = r_2$. ∎

It follows that every integer is congruent modulo $n$ to exactly one of the integers 0, 1, 2, ... , $n-1$, because these are the only possible remainders on dividing an integer by $n$.

## ■ *Example 4.4.2*

In the case $n = 2$, every integer is congruent to either 0 or 1, according to whether it is even or odd. Thus the integers $\mathbf{Z}$ are split into two sets: the even numbers, $2\mathbf{Z}$, and the odd numbers, $2\mathbf{Z}+1$.

## ■ *Example 4.4.3*

In the case $n = 3$, every integer is congruent to 0, 1 or 2. The integers $\mathbf{Z}$ split into three disjoint sets: $3\mathbf{Z}$, $3\mathbf{Z}+1$ and $3\mathbf{Z}+2$, where

$$3\mathbf{Z} = \{\ldots, -6, -3, 0, 3, 6, \ldots\}$$
$$3\mathbf{Z}+1 = \{\ldots, -5, -2, 1, 4, 7, \ldots\}$$
$$3\mathbf{Z}+2 = \{\ldots, -4, -1, 2, 5, 8, \ldots\}$$

In general, for any positive integer $n$, $\mathbf{Z}$ splits into $n$ disjoint sets, $n\mathbf{Z}$, $n\mathbf{Z}+1$, ... , $n\mathbf{Z}+(n-1)$, illustrated in Fig. 4.1.

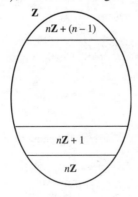

**Fig. 4.1**

**Definition**: For any $a \in \mathbf{Z}$, define the notation $[a]_n = n\mathbf{Z}+a$. Then the set $[a]_n = \{\ldots, a-2n, a-n, a, a+n, a+2n, \ldots\}$ is called the **residue class of $a$ modulo $n$**. Where there is no ambiguity the suffix $n$ will usually be dropped and $[a]$ will be used instead.

From the remark following Theorem 7, there are exactly $n$ residue classes of integers modulo $n$, namely $[0]$, $[1]$, $[2]$, ... , $[n-1]$. Let $\mathbf{Z}_n$ denote this set of residue classes.

In a number of ways, these residue classes behave like integers. You can add them and multiply them in ways justified by Theorem 8.

**Theorem 8**: Let $a \equiv b \,(\mathrm{mod}\, n)$ and $c \equiv d \,(\mathrm{mod}\, n)$. Then $a + c \equiv b + d \,(\mathrm{mod}\, n)$ and $ac \equiv bd \,(\mathrm{mod}\, n)$.

**Proof**: Suppose that $a \equiv b \,(\mathrm{mod}\, n)$ and $c \equiv d \,(\mathrm{mod}\, n)$. Then there exist integers $h$ and $k$ such that $a - b = hn$ and $c - d = kn$, or $a = b + hn$ and $c = d + kn$.

Add these equations: $a + c = (b + hn) + (d + kn) = b + d + n(h + k)$. This shows that $a + c \equiv b + d \,(\mathrm{mod}\, n)$.

From multiplying the equations, it follows that

$$ac = (b + hn)(d + kn)$$
$$= bd + hdn + bkn + hkn^2$$
$$= bd + n(hd + bk + hkn)$$

which shows that $ac \equiv bd \,(\mathrm{mod}\, n)$. ∎

Hence $[a] + [b] = [a + b]$ and $[a][b] = [ab]$ are both well-defined binary operations on $\mathbf{Z}_n$. They are called respectively, **addition modulo $n$** and **multiplication modulo $n$**.

## ■ *Example 4.4.3*

The table for $\mathbf{Z}_2$ using the operation $+$ is given by Fig. 4.2 and that for $\mathbf{Z}_3$ by Fig. 4.3.

|   | 0 | 1 |
|---|---|---|
| 0 | 0 | 1 |
| 1 | 1 | 0 |

|   | 0 | 1 | 2 |
|---|---|---|---|
| 0 | 0 | 1 | 2 |
| 1 | 1 | 2 | 0 |
| 2 | 2 | 0 | 1 |

**Fig. 4.2**                    **Fig. 4.3**

*Notice that the [ ] has been dropped. Strictly, [a] is both the set nℤ + a and also an element of ℤ$_n$. But in practice, provided that it is clear that you are working in ℤ$_n$, there is no confusion about dropping the bracket notation. This is another example of the situation referred to in Example 2.7.1, where you have a composite entity, (this time the set [a]) which should sometimes be thought of as a single compressed item, and at other times as a collection of pieces.*

### ■ *Example 4.4.4*

When you come to carry out calculations in $\mathbf{Z}_n$, it is probably easiest to calculate the remainder when you divide by $n$.

So, for example, in $\mathbf{Z}_8$, $3+4=7$ and $3\times 4 = 4$. Similarly, in $\mathbf{Z}_9$, $6+8=5$ and $6\times 8 = 3$.

## 4.5 SOME REMARKS

$\mathbf{Z}_n$ behaves like $\mathbf{Z}$ in many ways: for example, $a+b=b+a$, $a+0=a$, $a(b+c)=ab+ac$, $a.1=a$ and $a.0=0$.

However, one crucial difference is that in $\mathbf{Z}_n$ two non-zero numbers can be multiplied together to give zero. For example, in $\mathbf{Z}_4$, $2\times 2 = 0$, and in $\mathbf{Z}_6$, $2\times 3 = 0$.

However, this can never happen if $n$ is prime.

**Theorem 9**: Let $p$ be a prime number, and let $[a]$, $[b]\in\mathbf{Z}_p$. If $[a][b]=[0]$, then either $[a]=[0]$ or $[b]=[0]$.

**Proof**: If $[a][b]=[0]$ in $\mathbf{Z}_p$, then $[ab]=[0]$, giving $ab\equiv 0\,(\mathrm{mod}\,p)$. Therefore $p$ divides $ab$. By Theorem 4, either $p$ divides $a$, or $p$ divides $b$. Hence $a\equiv 0\,(\mathrm{mod}\,p)$ or $b\equiv 0\,(\mathrm{mod}\,p)$, so $[a]=[0]$ or $[b]=[0]$. ■

Another difference between $\mathbf{Z}_n$ and $\mathbf{Z}$ (in general) is that in $\mathbf{Z}_n$ the number of roots of a polynomial equation $f(x)=0$ can sometimes be greater than the degree of the equation.

For example, in $\mathbf{Z}_8$, the polynomial equation $x^2 -1 = 0$ has four roots, namely, 1, 3, 5 and 7.

However, this too can never happen if $n$ is prime.

**Theorem 10**: If $f(x)$ is a polynomial with coefficients in $\mathbf{Z}_p$, where $p$ is prime, then the number of elements $\alpha$ in $\mathbf{Z}_p$ for which $f(\alpha) = 0$ is less than, or equal to, the degree of $f(x)$.

**Proof**: The proof will not be given here. It uses Theorem 9 (which says that when you multiply two non-zero elements of $\mathbf{Z}_p$, the answer is non-zero) and is very similar to the proof of the corresponding result for polynomials with real coefficients. ∎

## WHAT YOU SHOULD KNOW

■ How to carry out calculations in $\mathbf{Z}_n$.

## EXERCISE 4

**1**  Find a counterexample to show that the result in Theorem 2 is not true if $m$ and $n$ are not relatively prime.

**2**  Find a counterexample to show that the result in Theorem 3 is not true if $m$ and $n$ are not relatively prime.

**3**  Which of the following statements are true and which are false?

(1)   $7 \equiv 9 \pmod 2$

(2)   $3 \equiv -5 \pmod 4$

(3)   $5 \not\equiv -13 \pmod 3$

**4**  Carry out each of the following calculations.

(1)   $2 + 2$ in $\mathbf{Z}_3$

(2)   $3 + 5$ in $\mathbf{Z}_8$

(3)   $7 \times 9$ in $\mathbf{Z}_{10}$

(4)   $6 \times 2$ in $\mathbf{Z}_{12}$

**5**  Solve the equation $x^2 \equiv 3 \pmod{11}$.

**6**  Let $a$ and $b$ be any two positive integers, and let $h$ be their highest common factor and $l$ be their least common multiple. Prove that $ab = hl$. (Hint: Observe that if $a = p_1^{r_1} \ldots p_n^{r_n}$ and $b = p_1^{s_1} \ldots p_n^{s_n}$ then $h = p_1^{\alpha_1} \ldots p_n^{\alpha_n}$ and $l = p_1^{\beta_1} \ldots p_n^{\beta_n}$ where $\alpha_i = \min\{r_i, s_i\}$, $i = 1, \ldots, n$ and $\beta_i = \max\{r_i, s_i\}$, $i = 1, \ldots, n$.)

# 5

# *Groups*

## 5.1 INTRODUCTION

This chapter begins a sequence of chapters which look more carefully at structure within sets, and analyse it in detail. You will also see that some apparently different sets and structures share some similar features.

## 5.2 TWO EXAMPLES OF GROUPS

### ■ *Example 5.2.1*

In Section 3.4 you considered the set $X = \{2, 4, 6, 8\}$ under the operation of multiplying and then taking the remainder after dividing by 10. You were able to construct the table in Fig 5.1, which is a copy of Fig. 3.1.

|  | | Second number | | | |
|---|---|---|---|---|---|
|  | ○ | 2 | 4 | 6 | 8 |
|  | 2 | 4 | 8 | 2 | 6 |
| First | 4 | 8 | 6 | 4 | 2 |
| number | 6 | 2 | 4 | 6 | 8 |
|  | 8 | 6 | 2 | 8 | 4 |

Fig. 5.1

If you look at Fig. 5.1 closely, you will see two features.

● One row is exactly the same as the top row, and one column is exactly the same as the left-hand column.

● Each row and each column consists of exactly the elements $\{2,4,6,8\}$ in some order, each element occurring just once.

These two features are among the characteristics of a group.

## ■ *Example 5.2.2*

Suppose that ABC is an equilateral triangle, called **T**. Consider the following six transformations of the plane containing **T**.

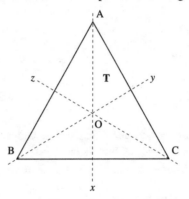

**Fig. 5.2**

● *X* means     'Reflect in the line *x*.'
● *Y* means     'Reflect in the line *y*.'
● *Z* means     'Reflect in the line *z*.'
● *R* means     'Rotate by 120° anticlockwise about O.'
● *S* means     'Rotate by 240° anticlockwise about O.'
● *I* means     'Do nothing.'

You can see that each of these transformations leaves **T** in the same overall position, even though it may change the positions of the points which make up **T**. Such a transformation is called a **symmetry** of **T**. Thus the transformation *R* puts A in the position occupied by B, B in the position occupied by C, and C in the position occupied by A. The transformation *X* interchanges B and C and leaves A where it is.

To combine operations, you use the rule 'followed by'. When you carry out the transformation $XR$, this is $R$ followed by $X$, A goes to the position occupied by B and then to the position initially occupied by C, B goes to the position occupied by C and then back to the position where it was at the start, and C goes to the position occupied by A. The result of this combined transformation is the same as the transformation carried out by $Y$. So you write $XR = Y$. Similarly $RS = I$.

> *You may find it useful to make a triangle out of card and write A, B and C on both faces. You can then work out the results quickly.*

In a similar way, each of the transformations $I, R, S, X, Y, Z$ can be combined with another one, or itself, using the rule 'followed by' and the result is always again one of these same six transformations. Hence 'followed by' is a binary operation on the set $\{I, R, S, X, Y, Z\}$.

Fig. 5.3 shows the table for this binary operation.

|   | $I$ | $R$ | $S$ | $X$ | $Y$ | $Z$ |
|---|-----|-----|-----|-----|-----|-----|
| $I$ | $I$ | $R$ | $S$ | $X$ | $Y$ | $Z$ |
| $R$ | $R$ | $S$ | $I$ | $Z$ | $X$ | $Y$ |
| $S$ | $S$ | $I$ | $R$ | $Y$ | $Z$ | $X$ |
| $X$ | $X$ | $Y$ | $Z$ | $I$ | $R$ | $S$ |
| $Y$ | $Y$ | $Z$ | $X$ | $S$ | $I$ | $R$ |
| $Z$ | $Z$ | $X$ | $Y$ | $R$ | $S$ | $I$ |

**Fig. 5.3**

Once again, notice that:
- one row is exactly the same as the top row, and one column is exactly the same as the left-hand column
- each row and each column consists of exactly the elements $\{I, R, S, X, Y, Z\}$ in some order, each element occurring just once.

There is one other property shared by the tables in Figs. 5.1 and 5.3. This is the fact that if you multiply three elements $a$, $b$ and $c$, then it does not matter whether you calculate first $ab$ and then $(ab)c$, or first $bc$ and then $a(bc)$.

As you will see in the next section, the sets in Examples 5.2.1 and 5.2.2, together with the binary operations, form groups.

The group of transformations described in Example 5.2.2 is one of a family of groups called dihedral groups (discussed in Chapter 13). This group is denoted by $D_3$: $D$ is for dihedral and the 3 indicates that the group concerns the symmetries of an equilateral triangle.

## 5.3 DEFINITION OF A GROUP

**Definition**: A **group** is a set $G$, with a binary operation $\circ$, such that:

(1)   $x \circ y \in G$ for all $x, y \in G$; that is, the operation $\circ$ is closed.

(2)   $x \circ (y \circ z) = (x \circ y) \circ z$ for all $x, y, z \in G$; that is the operation is associative.

(3)   There exists an element $e \in G$ such that for all $x \in G$, $e \circ x = x \circ e = x$.

(4)   For each element $x \in G$, there exists an element $x^{-1} \in G$ such that $x^{-1} \circ x = x \circ x^{-1} = e$.

The group $G$ with operation $\circ$ is written $(G, \circ)$.

> *Notice that the first requirement is already implicit in the term 'binary operation'. The reason for stating it separately is that it is helpful, when proving that a set G with an operation $\circ$ is a group, to realise that you need to prove four things.*

The element $e$ described in the third condition is called the **identity** for the group $(G, \circ)$.

> *You need to be careful, because you cannot say it is **the** identity element until you have proved that there can be only one. The proof starts by supposing that there are two of them.*

**Theorem 11**: The identity element of a group $(G, \circ)$ is unique.

**Proof**: Suppose that there are two elements of $(G, \circ)$ with property (3). Call them $e$ and $e'$. Then, consider $e \circ e'$. Using the fact that $e$ is the identity, $e \circ e' = e'$. Similarly, using the fact that $e'$ is the identity, $e \circ e' = e$. Therefore $e = e'$, and the identity element $e$ is unique. ∎

The element $x^{-1}$ described in the fourth condition for a group is called the **inverse** of $x$. Once again this element is unique for a given element $x$. That is, a given element $x$ cannot have two inverses.

**Theorem 12**: Given a group $(G, \circ)$, and an element $x \in G$, there is only one element $y \in G$ such that $y \circ x = x \circ y = e$.

**Proof**: Suppose that there are two elements $y$ and $y'$ with the property that $y \circ x = x \circ y = e$ and $y' \circ x = x \circ y' = e$. Then

$$
\begin{aligned}
y' &= y' \circ e & & e \text{ is the identity} \\
&= y' \circ (x \circ y) & & \text{since } x \circ y = e \\
&= (y' \circ x) \circ y & & \text{the associativity property} \\
&= e \circ y & & \text{since } y' \circ x = e \\
&= y & & e \text{ is the identity}
\end{aligned}
$$

Therefore $y = y'$, and the inverse of $x$ is unique. ∎

*In Example 5.2.1, the identity element is 6; in Example 5.2.2, $D_5$, it is $I$. You can check from the tables in Figs. 5.1 and 5.2 that each element has an inverse. The associativity of the binary operation in Example 5.2.1 follows from the associativity of multiplication of the integers. In Example 5.2.2, to see that the associative property of transformations under 'followed by' is true, note that doing the transformation $(ab)c$ is equivalent to doing $c$, then $b$, then $a$, as is doing the transformation $a(bc)$. This is shown more formally in Theorem 32. So the sets and their binary operations in Examples 5.2.1 and 5.2.2 are both groups.*

**Definition**: A group $G$ is said to be **abelian** if it also satisfies $x \circ y = y \circ x$ for all $x, y \in G$; that is, if the operation is commutative.

*Abelian groups are so named after the mathematician Abel (1802–1829) who pioneered early work on such groups.*

## 5.4 A DIVERSION ON NOTATION

It can be tedious always to talk about the group $(G, \circ)$, and it is quite usual, and quite acceptable to leave out the operation and to use multiplicative notation. So, provided there is no ambiguity, the notation $G$ on its own will be used for a group, with the understanding that $xy$ will be used instead of $x \circ y$.

Unfortunately, there is an exception to this. There are some groups, such as the integers under addition, in which it is usual to denote the binary operation by +. In this case, it is more convenient to use additive notation for the group operation.

In additive notation, a group is a set $G$ with a binary operation + satisfying:
(1)  $x + y \in G$ for all $x, y \in G$.
(2)  $x + (y + z) = (x + y) + z$ for all $x, y, z \in G$.
(3)  There exists an element $0 \in G$ such that for all $x \in G$, $0 + x = x + 0 = x$.
(4)  For each element $x \in G$, there exists an element $-x \in G$ such that $(-x) + x = x + (-x) = 0$.

From now on multiplicative notation will be used in this book except where it is more natural to use additive notation.

> You may very well think that it is bad not to be consistent with notation, but just think of the problems working in the group $(\mathbf{Z}, +)$ with multiplicative notation. You would be writing $3.5 = 8$, which could raise more problems than a little inconsistency of notation.

In practice, this inconsistency of notation should not be a problem.

Note that it is not usual to use additive notation if the group is not abelian.

## 5.5 SOME EXAMPLES OF GROUPS

### ▉ *Example 5.5.1*

The non-zero rational numbers $\mathbf{Q}*$ under multiplication form a group which is written as $(\mathbf{Q}*, \times)$. Similarly, $(\mathbf{R}*, \times)$ and $(\mathbf{C}*, \times)$ are also groups. In each of them the identity element is 1.

### ▉ *Example 5.5.2*

As the tables in Fig. 4.2 and Fig. 4.3 might suggest to you, $\mathbf{Z}_n$ with the operation of addition forms a group $(\mathbf{Z}_n, +)$.

**Theorem 13**: $(\mathbf{Z}_n, +)$ is a group.

**Proof:** The remark in Section 4.4 after Theorem 8 shows that $[a] + [b] = [a + b]$ is a well-defined binary operation on $\mathbf{Z}_n$.

To show that the operation + is associative, note that:

$$[a] + ([b] + [c]) = [a] + [b + c]$$
$$= [a + (b + c)]$$
$$= [(a + b) + c]$$
$$= [a + b] + [c]$$
$$= ([a] + [b]) + [c]$$

> The step $[a + (b + c)] = [(a + b) + c]$ above follows because addition of integers is associative. In this sense, the associativity of $(\mathbf{Z}_n, +)$ is 'inherited' from the associativity of $(\mathbf{Z}, +)$.

The element $[0]$ is the identity element for $(\mathbf{Z}_n, +)$ as, for $[a] \in \mathbf{Z}_n$, $[0] + [a] = [0 + a] = [a]$ and $[a] + [0] = [a + 0] = [a]$.

Finally, for $[a] \in \mathbf{Z}_n$, the inverse is $[-a] \in \mathbf{Z}_n$ because $[a] + [-a]$ $= [a + (-a)] = [0]$ and $[-a] + [a] = [(-a) + a] = [0]$.

Therefore $\mathbf{Z}_n$, together with the operation +, is a group. ▉

This group, which strictly should be written as $(\mathbf{Z}_n, +)$, will sometimes be written simply $\mathbf{Z}_n$ to avoid cumbersome notation.

## ■ *Example 5.5.3*

The integers $\mathbf{Z}$ under addition form a group, $(\mathbf{Z},+)$, in which the identity element is the integer 0.

## ■ *Example 5.5.4*

The rotations of a circle about its centre form a group under the operation 'followed by'. The identity element is the 'leave it alone' rotation of $0°$.

## ■ *Example 5.5.5*

Consider the set $\mathbf{Z}_5{}^*$, with the operation of multiplication. Its table is shown in Fig. 5.3.

|   | 1 | 2 | 3 | 4 |
|---|---|---|---|---|
| 1 | 1 | 2 | 3 | 4 |
| 2 | 2 | 4 | 1 | 3 |
| 3 | 3 | 1 | 4 | 2 |
| 4 | 4 | 3 | 2 | 1 |

**Fig. 5.3**

You can easily verify that this is a group table.

More generally, let $p$ be a prime number. Then the set $\mathbf{Z}_p{}^*$ is a group under multiplication.

**Theorem 14**: Let $p$ be a prime number. Then the set $\mathbf{Z}_p{}^*$ with the operation of multiplication is a group, $\left(\mathbf{Z}_p{}^*,\times\right)$.

*From Section 4.4, multiplication is a binary operation on $\mathbf{Z}_p$. But is it a binary operation on $\mathbf{Z}_p{}^*$? You need to check that when two non-zero elements of $\mathbf{Z}_p$ are multiplied together, the result is also non-zero.*

If $[a],[b] \in \mathbf{Z}_p{}^*$, then $[a] \neq [0]$ and $[b] \neq [0]$. So by Theorem 9, $[a][b] \neq [0]$, and hence $[a],[b] \in \mathbf{Z}_p{}^*$.

To check that the operation of multiplication is associative, note that

$$[a]([b][c]) = [a][bc]$$
$$= [a(bc)]$$
$$= [(ab)c]$$
$$= [ab][c]$$
$$= ([a][b])[c]$$

The identity element is $[1]$, because, for any $[a] \in \mathbf{Z}_p{}^*$, $[a][1] = [a1] = [a]$ and $[1][a] = [1a] = [a]$.

Finally, suppose that $[a] \in \mathbf{Z}_p{}^*$. Then $[a] \neq [0]$ and so $p$ is not a factor of $a$. As the only factors of $p$ are 1 and $p$, this means that $a$ and $p$ are relatively prime. Therefore, by Theorem 1, there exist integers $x$ and $y$ such that $ax + py = 1$. It follows that $[a][x] = [x][a] = [1]$, showing that $[x]$ is an inverse for the element $[a]$. So each element of $\mathbf{Z}_p{}^*$ has an inverse. Therefore $\left( \mathbf{Z}_p{}^*, \times \right)$ is a group. ∎

All the examples of groups in this section so far have been abelian groups. Here is an example of a non-abelian group.

## ■ *Example 5.5.6*

Let **M** be the set of invertible $2 \times 2$ matrices with real entries, together with the operation of matrix multiplication. Then $(\mathbf{M}, \times)$ is a group.

Note that this group is not abelian since

$$\begin{pmatrix} 1 & 1 \\ 0 & 1 \end{pmatrix}\begin{pmatrix} 1 & 1 \\ 0 & 2 \end{pmatrix} = \begin{pmatrix} 1 & 3 \\ 0 & 2 \end{pmatrix} \text{ and } \begin{pmatrix} 1 & 1 \\ 0 & 2 \end{pmatrix}\begin{pmatrix} 1 & 1 \\ 0 & 1 \end{pmatrix} = \begin{pmatrix} 1 & 2 \\ 0 & 2 \end{pmatrix}.$$

Looking back at these examples, you can see that they have different numbers of elements.

For example, the groups in Examples 5.5.1, 5.5.3, 5.5.4 and 5.5.6 each have an infinite number of elements, in Example 5.5.2, $(\mathbf{Z}_n, +)$ has $n$ elements, and in Example 5.5.5, $\left( \mathbf{Z}_p{}^*, \times \right)$ has $p - 1$ elements.

**Definition**: The number of elements of a finite group is called the **order** of the group. If the group has an infinite number of elements it is said to have **infinite order**.

## 5.6 SOME USEFUL PROPERTIES OF GROUPS

Here are some elementary properties of groups which will be used throughout the rest of the book. For convenience the properties are stated as a theorem.

**Theorem 15**: Let $G$ be a group. Then the following properties are true.

(1)   For $a, b \in G$, if $ab = e$, then $a = b^{-1}$ and $b = a^{-1}$.

(2)   $(ab)^{-1} = b^{-1}a^{-1}$ for all $a, b \in G$.

(3)   $\left(a^{-1}\right)^{-1} = a$ for all $a \in G$.

(4)   For $a, x, y \in G$, if $ax = ay$ then $x = y$ and if $xa = ya$ then $x = y$.

**Proof**:

> *In following the proofs of these results, look back at the group axioms to find the appropriate reason for each step.*

(1)   Let $ab = e$. For the first part, multiply on the right by $b^{-1}$. Then $(ab)b^{-1} = eb^{-1}$, so $a\left(bb^{-1}\right) = b^{-1}$, leading to $ae = b^{-1}$ and $a = b^{-1}$. For the second part multiply $ab = e$ on the left by $a^{-1}$. Then $a^{-1}(ab) = a^{-1}e$, so $\left(a^{-1}a\right)b = a^{-1}$, leading to $eb = a^{-1}$ and $b = a^{-1}$.

(2)   Take the expression $(ab)\left(b^{-1}a^{-1}\right)$ and prove that it is equal to $e$. Thus $(ab)\left(b^{-1}a^{-1}\right) = a\left(b\left(b^{-1}a^{-1}\right)\right) = a\left(\left(bb^{-1}\right)a^{-1}\right) = a\left(ea^{-1}\right) = aa^{-1} = e$. Then, from part (1), $b^{-1}a^{-1}$ is the inverse of $ab$.

(3)   From the expression $aa^{-1} = e$, and using part (1) with $b$ as $a^{-1}$, the inverse of $b$ is $a$, so the inverse of $a^{-1}$ is $a$. Therefore $\left(a^{-1}\right)^{-1} = a$.

(4)   If $ax = ay$, then $a^{-1}(ax) = a^{-1}(ay)$, so $\left(a^{-1}a\right)x = \left(a^{-1}a\right)y$ leading to $ex = ey$ and $x = y$. Similarly, if $xa = ya$, then $(xa)a^{-1} = (ya)a^{-1}$, so $x\left(aa^{-1}\right) = y\left(aa^{-1}\right)$, leading to $xe = ye$ and $x = y$.

> *The results in part (4) are sometimes called the cancellation laws for a group.*

## 5.7 THE POWERS OF AN ELEMENT

When you combine an element $x$ of a group $G$ with itself, you obtain $xx$ which it is natural to write as $x^2$. This makes it useful to define exactly what is meant by powers of an element.

**Definition**: Let $x$ be an element of a group $G$. Then if $s$ is a positive integer, $x^s = \overbrace{xx\ldots x}^{s \text{ times}}$ and $x^{-s} = \overbrace{x^{-1}x^{-1}\ldots x^{-1}}^{s \text{ times}}$. The power $x^0$ is defined to be $x^0 = e$.

It is not difficult to prove all the usual rules of indices from this definition. Thus $x^s x^t = x^{s+t}$, $\left(x^s\right)^t = x^{st}$ and $x^{-s} = \left(x^s\right)^{-1} = \left(x^{-1}\right)^s$ for all $s, t \in \mathbf{Z}$.

*These rules are not proved here but the proofs are left as an exercise.*

Consider the group $D_3$ in Example 5.2.2, for which the multiplication table is shown in Fig. 5.4. Look at the subset $H$ of this group consisting of powers of the element $S$.

|   | I | R | S | X | Y | Z |
|---|---|---|---|---|---|---|
| I | I | R | S | X | Y | Z |
| R | R | S | I | Z | X | Y |
| S | S | I | R | Y | Z | X |
| X | X | Y | Z | I | R | S |
| Y | Y | Z | X | S | I | R |
| Z | Z | X | Y | R | S | I |

**Fig. 5.4** *The group* $D_3$

You can see that $S^2 = S \circ S = R$. Also $S^3 = S \circ S^2 = S \circ R = I$. Any positive power $n$ of $S$ will now be either $S$, $R$, or $I$ depending on the remainder when $n$ is divided by 3.

Also, as $S^3 = I$, $S^{-1} = S^2$, so every negative power of $S$ is also a positive power of $S$.

The only distinct powers of $S$ are thus $S$, $S^2$ and $S^3$, which is equal to the identity.

A similar situation holds in any finite group.

**Theorem 16**: Let $x$ be an element of a finite group $G$. Then the powers of $x$ cannot all be different, and there is a smallest positive integer $k$ such that $x^k = e$.

**Proof**: Consider all positive powers of $x$. These powers cannot all be different, because the group is finite, so at some stage two elements must be the same. Suppose that two of these elements are $x^r$ and $x^s$, with $r \leq s$ such that $x^r = x^s$. Then $x^{s-r} = x^s x^{-r} = x^r x^{-r} = x^r \left(x^r\right)^{-1} = e$, so that $x^{s-r} = e$. There is thus a positive power of $x$ which gives the identity element. By the well-ordering principle, there is a smallest such power, say $k$; then $x^k = e$. ∎

The result of Theorem 16 leads to an important definition.

## 5.8 THE ORDER OF AN ELEMENT

**Definition**: An element $x$ of a group $G$ is said to have **finite order** if $x^n = e$ for some $n > 0$; the least such $n$ is called the **order** of $x$. If no such $n$ exists, $x$ is said to have **infinite order**.

Some theorems follow immediately from the definition.

**Theorem 17**: Let $G$ be a group.

(1)    Let $x \in G$ be an element of infinite order, and let $N$ be an integer. Then $x^N = e$ if, and only if, $N = 0$.

(2)    Let $x \in G$ be an element of finite order $n$, and let $N$ be an integer. Then $x^N = e$ if, and only if, $n$ divides $N$.

**Proof:** (1) *If.* If $N = 0$, then $x^N = e$, by the definition of powers.

*Only if.* Suppose that $x^N = e$. From the definition of infinite order, you cannot have $N > 0$. But if $x^N = e$, then $x^{-N} = \left(x^N\right)^{-1} = e^{-1} = e$, so you cannot have $N < 0$ either. Therefore $N = 0$.

(2) *If.* Suppose that $n$ divides $N$, and that $N = kn$ for an integer $k$. Then $x^N = x^{kn} = \left(x^n\right)^k = e^k = e$.

*Only if.* Suppose that $x^N = e$. Then, using the division algorithm, $N = qn + r$ where $0 \le r < n$, so $e = x^N = x^{qn+r} = \left(x^n\right)^q x^r = e^q x^r = x^r$. Therefore $r = 0$, because $n$ is the least positive power of $x$ such that $x^n = e$. Therefore $N = qn$, and so $n$ divides $N$. ∎

**Theorem 18**: Let $G$ be a group.

(1) If an element $x$ has infinite order, then all the powers of $x$ are distinct.

(2) If an element $x$ has finite order $n$, then $x^r = x^s$ if and only if $r \equiv s \pmod{n}$, so the powers of $x$ repeat in cycles of length $n$.

**Proof**: (1) If $x^r = x^s$, then $x^{r-s} = x^r x^{-s} = x^s x^{-s} = x^s \left(x^s\right)^{-1} = e$. By part (1) of Theorem 17, $r - s = 0$ and so $r = s$. Therefore, if $r \ne s$, $x^r \ne x^s$.

(2) *If.* If $r \equiv s \pmod{n}$, then $n$ divides $r - s$, so $r - s = kn$ for some integer $k$. Therefore $x^r = x^{s+kn} = x^s x^{kn} = x^s \left(x^n\right)^k = x^s e^k = x^s$.

*Only if.* If $x^r = x^s$, then $x^{r-s} = e$, as in part (1), and by part (2) of Theorem 17, $n$ divides $r - s$. So $r \equiv s \pmod{n}$.

Therefore if $0 \le r < s \le n-1$, then $x^r \ne x^s$, so the elements $e, x, x^2, \ldots, x^{n-1}$ are all distinct. But any power of $x$ is equal to one of these elements. For consider $x^m$, where $m$ is an integer. By the division algorithm, $m = qn + r$ where $0 \le r < n$. It follows that $x^m = x^{qn+r} = \left(x^n\right)^q x^r = e^q x^r = x^r$ where $0 \le r < n$. Therefore the powers repeat themselves in cycles, and the cycles are of length $n$. ∎

There are two more theorems about orders of elements which will be needed for Section 14.5.

*You may wish to leave these theorems until you study Chapter 14.*

**Theorem 19**: Let $x$ be an element of a group $G$, and let the order of $x$ be $n$.

(1)    If $m$ and $n$ are relatively prime, then $x^m$ has order $n$.

(2)    If $n = st$, then $x^s$ has order $t$.

**Proof**: (1) First note that $\left(x^m\right)^n = x^{mn} = \left(x^n\right)^m = e^m = e$. Suppose that $\left(x^m\right)^d = e$. Then $x^{md} = e$. Hence, by Theorem 17, part (2), $n$ divides $md$. But $m$ and $n$ are relatively prime. Therefore, by Theorem 2, $n$ divides $d$. Therefore $n$ is the least positive integer satisfying $\left(x^m\right)^n = e$. Therefore the order of $x^m$ is $n$.

(2) $\left(x^s\right)^t = x^{st} = x^n = e$. Now let $d$ be any positive integer satisfying $\left(x^s\right)^d = e$. Then $x^{sd} = e$. Therefore $n$ divides $sd$, by Theorem 17, part (2). Therefore $st$ divides $sd$, so $t$ divides $d$. Therefore $t$ is the least power of $x^s$ to give the identity. Therefore the order of $x^s$ is $t$. ∎

**Theorem 20**: Let $G$ be an abelian group and let $a$ and $b$ be elements of $G$ having orders $m$ and $n$ respectively.

(1)    If $m$ and $n$ are relatively prime, then $ab$ has order $mn$.

(2)    Let $l$ be the least common multiple of $m$ and $n$. Then there exists an element of $G$ having order $l$.

**Proof**: (1) Let the order of $ab$ be $k$. Then, since $G$ is abelian, $(ab)^n = a^n b^n$, and as $n$ is the order of $b$, $(ab)^n = a^n b^n = a^n$ which has order $m$, by Theorem 19, part (1). But $\left((ab)^n\right)^k = \left((ab)^k\right)^n = e^n = e$, so $m$ divides $k$, by Theorem 17, part (2). Similarly, $n$ divides $k$. Therefore, by Theorem 3, $mn$ divides $k$, since $m$ and $n$ are relatively prime. But $(ab)^{mn} = \left(a^m\right)^n \left(b^n\right)^m = e^n e^m = e$. Therefore $k = mn$.

(2) Let $h$ be the highest common factor of $m$ and $n$. Then $h$ divides $m$ and $h$ divides $n$, so there exist integers $s$ and $t$ such that $m = sh$ and $n = th$. Moreover, $s$ and $t$ are relatively prime. If they were not, they would have a common factor, $k > 1$, which would mean that $hk$ is a number greater than $h$ which divides $m$ and $n$, thus contradicting the fact that $h$ is the highest common factor. Now, $a^h$ has order $s$, by Theorem 19, part (2). Also, since $s$ and $n$ are relatively prime, (because $m$ and $n$ are relatively prime and $s$ divides $m$), by what has just been proved in part (1), $a^h b$ has order $sn$. But, by Theorem 6, $mn = hl$, where $l$ is the

least common multiple of $m$ and $n$. Therefore $hsn = mn = hl$. Therefore $sn = l$, so $a^h b$ is an element of order $l$. ∎

## WHAT YOU SHOULD KNOW

■ The definition of a group.

■ That the identity element for a group is unique.

■ That each element in a group has a unique inverse element.

■ How to write out a group table for a small finite group.

■ That multiplicative notation is generally used for groups.

■ What is meant by the order of a group.

■ The meaning of the power of an element.

■ The meaning of the order of an element.

## EXERCISE 5

**1**  Write out a table showing the set $\{1, 5, 7, 11\}$ under the operation of multiplication modulo 12. Is it the table of a group?

**2**  The notation $\mathbf{R} - \{0, 1\}$ means the set $\mathbf{R}$ without the numbers 0 and 1. Consider the set of functions $\{I, F, G, H, K, L\}$ defined on the set $\mathbf{R} - \{0, 1\}$ defined by $I(x) = x$, $F(x) = 1/(1 - x)$, $G(x) = (x - 1)/x$, $H(x) = 1 - x$, $K(x) = x/(x - 1)$ and $L(x) = 1/x$. Construct a table to show that composition of functions is a binary operation on this set, and that the set $\{I, F, G, H, K, L\}$ together with the operation of composition is a group. (Assume that composition of functions is associative; this is actually proved in Chapter 10, Theorem 32.)

**3**  Show that the set $\mathbf{R} - \{-1\}$, together with the operation ∘, given by $x \circ y = x + y + xy$, is a group. Which is the identity element for the group? Find an expression for $x^{-1}$ in terms of $x$. (Note: the notation $\mathbf{R} - \{-1\}$ means the set $\mathbf{R}$ with the number $-1$ deleted.)

**4**  Let R be a rectangle which is not a square. Follow the working of Example 5.2.2, and define a set of symmetries for R. Draw up a group table for these symmetries. This group is called the Vierergruppe, the four-group, and denoted by $V$.

**5**  Write out a table for the set $\{1,3,7,9\}$ under the operation multiplication modulo 10.

**6**  Write out a table for the set $\{1,2,4,8\}$ under the operation multiplication modulo 15. Show that the table has the same structure as the table in question 5, with the elements re-named.

**7**  Let $G$ be a group, and let $a$ be an element of $G$. Prove that if $a^2 = a$, then $a = e$.

**8**  Let $G$ be a group. Prove that if $a^2 = e$ for all $a \in G$, then $G$ is abelian.

**9**  Which of the following are groups? Give reasons for your answer in each case.

(1)  The set of all odd integers under addition.

(2)  The set of all integers of the form $2^m 5^n$ $(m,n \in \mathbf{Z})$, under multiplication.

(3)  The set of all $2 \times 2$ matrices of the form $\begin{pmatrix} a & b \\ 0 & 0 \end{pmatrix}$, where $a,b \in \mathbf{R}$, under matrix multiplication.

(4)  The set of all $2 \times 2$ matrices of the form $\begin{pmatrix} a & 0 \\ b & c \end{pmatrix}$, where $a,c \in \mathbf{R}$, $a \neq 0$, $c \neq 0$, and $b \in \mathbf{Q}$ under matrix multiplication.

**10**  Show that the matrices $\left\{ \begin{pmatrix} 1 & 0 \\ 0 & 1 \end{pmatrix}, \begin{pmatrix} -1 & 0 \\ 0 & 1 \end{pmatrix}, \begin{pmatrix} 1 & 0 \\ 0 & -1 \end{pmatrix}, \begin{pmatrix} -1 & 0 \\ 0 & -1 \end{pmatrix} \right\}$ form a group under matrix multiplication.

**11**  Prove the three index rules $x^s x^t = x^{s+t}$, $\left( x^s \right)^t = x^{st}$ and $x^{-s} = \left( x^s \right)^{-1} = \left( x^{-1} \right)^s$ for all $s,t \in \mathbf{Z}$.

**12**  Mark each of the following statements true or false.

(1)  Every element of a group has an inverse.

(2)  It is possible for an element of a group to be its own inverse.

(3)  It is not possible for a group to be infinite.

(4)  It is not possible for a group to consist of a single element.

(5)  No element of a group has order 1.

(6)  If the order of an element $x$ is $n$ and $x^N = e$, then $N$ divides $n$.

(7)  All groups are abelian.

**13**  Prove the following generalisation of Theorem 19. Let $x$ be an element of a group $G$, and let the order of $x$ be $n$. If $h$ is the highest common factor of $n$ and $s$, then $x^s$ has order $n/h$.

# 6

## Subgroups

## 6.1 SUBGROUPS

Sometimes, within a group table, you can see a group inside the main group. For example, Fig. 6.1 shows the group $D_3$, which you first saw in Example 5.2.2. The section which is shaded and which is reproduced unshaded in Fig. 6.2, is itself a group with the same operation as that of the main group.

|   | I | R | S | X | Y | Z |
|---|---|---|---|---|---|---|
| I | I | R | S | X | Y | Z |
| R | R | S | I | Z | X | Y |
| S | S | I | R | Y | Z | X |
| X | X | Y | Z | I | R | S |
| Y | Y | Z | X | S | I | R |
| Z | Z | X | Y | R | S | I |

Fig. 6.1

|     | $I$ | $R$ | $S$ |
|-----|-----|-----|-----|
| $I$ | $I$ | $R$ | $S$ |
| $R$ | $R$ | $S$ | $I$ |
| $S$ | $S$ | $I$ | $R$ |

**Fig. 6.2**

The group in Fig. 6.2 is called a subgroup of the original group.

This leads to the following definition.

**Definition**: If $H$ is a subset of a group $G$ with operation $\circ$, such that $H$ is a group with the operation $\circ$, then $H$ is a **subgroup** of $G$.

*This is a natural definition, and one that you would expect. A subgroup is a subset which is itself a group under the operation inherited from G. Section 6.2 gives some examples of subgroups, and Section 6.3 gives a test that you can apply to tell whether a subset of a group is a subgroup.*

## 6.2 EXAMPLES OF SUBGROUPS

### ■ *Example 6.2.1*

Every group $G$ has two trivial subgroups, the group consisting of the identity $\{e\}$ alone, and the whole group $G$. Subgroups other than the identity subgroup and the whole group are called **proper** subgroups.

### ■ *Example 6.2.2*

In the group $D_3$ in Example 5.2.2, the set $\{I, X\}$ is a proper subgroup of $D_3$. Similarly, $\{I, Y\}$ and $\{I, Z\}$ are also proper subgroups.

However, the subset $\{I, R\}$ is not a subgroup. If you check you find that $RR = S$, so the operation is not closed on $\{I, R\}$.

Similarly $\{I, X, S\}$ is not a subgroup because $XS = Z$, which is not an element of $\{I, X, S\}$.

## ■ *Example 6.2.3*

The non-zero rational numbers $\mathbf{Q}^*$ under multiplication form a group $(\mathbf{Q}^*, \times)$. The set of positive rationals $\mathbf{Q}^+$ under multiplication, $(\mathbf{Q}^+, \times)$, is a proper subgroup of $(\mathbf{Q}^*, \times)$.

## ■ *Example 6.2.4*

The rational numbers $\mathbf{Q}$ with the operation of addition form a group $(\mathbf{Q}, +)$. However, the non-zero rational numbers $\mathbf{Q}^*$ do not form a group under addition because addition is not closed; $1 \in \mathbf{Q}^*$ and $-1 \in \mathbf{Q}^*$ but $1 + (-1) \notin \mathbf{Q}^*$. Therefore $(\mathbf{Q}^*, +)$ is not a subgroup of $(\mathbf{Q}, +)$.

Note that $(\mathbf{Q}^*, \times)$ is a group, but this fact does not make it a subgroup of $(\mathbf{Q}, +)$ because the operations are different.

## ■ *Example 6.2.5*

The sets $\{1, -1\}$, $\{1, -1, i, -i\}$ and $\{z \in \mathbf{C} : |z| = 1\}$ are all subgroups of $(\mathbf{C}^*, \times)$. In fact, $\{1, -1\}$ is a subgroup of $\{1, -1, i, -i\}$, which is in turn a subgroup of $\{z \in \mathbf{C} : |z| = 1\}$.

## ■ *Example 6.2.6*

The subset of $(\mathbf{Z}, +)$ which consists of multiples of 2, that is, $\{\ldots, -4, -2, 0, 2, 4, \ldots\}$ is a subgroup of $\mathbf{Z}$. This group will be called $(2\mathbf{Z}, +)$.

Similarly, for a positive integer $n$ the set $n\mathbf{Z} = \{kn : k \in \mathbf{Z}\}$ with the operation addition is a subgroup of $(\mathbf{Z}, +)$. This group will be called $(n\mathbf{Z}, +)$.

## 6.3 TESTING FOR A SUBGROUP

Although it is straightforward in particular cases for small finite groups to tell whether a subset of a group is a subgroup, it can be more difficult with larger groups. You need a way of testing systematically whether a subset is a subgroup. Theorem 21 gives such a test.

**Theorem 21**: Let $G$ be a group, and $H$ be a subset of $G$. Then $H$ is a subgroup of $G$ if, and only if:

- $xy \in H$ for each $x \in H$, $y \in H$
- $e \in H$
- $x^{-1} \in H$ for each $x \in H$.

**Proof:**

*Remember that, a theorem involving 'if, and only if,' requires proving in two directions.*

*To prove the 'if' result, you need to suppose that the conditions given in the theorem hold.*

*To show that H is a subgroup of G you need to show that H, together with the operation of G, satisfies the four conditions in the definition of a group given in Section 5.3.*

*If.* The first condition, that $xy \in H$ for each $x \in H$, $y \in H$, makes sure that $H$ is closed under the operation in $G$.

Suppose that $x, y$ and $z \in H$. Then, as $H$ is a subset of $G$, $x, y$ and $z \in G$. But $x(yz) = (xy)z$, since $G$ is a group. Thus associativity in $H$ is inherited from associativity in $G$.

The element $e \in H$, and $ex = xe = x$ for all $x \in H$, because this is true for all $x \in G$, and every element of $H$ is an element of $G$.

Given $x \in H$, $x^{-1} \in H$ and then $x^{-1}x = xx^{-1} = e$, because $x^{-1}$ is the inverse of $x$ in $G$.

*Only if.* Suppose that $H$ is a subgroup of $G$.

The first condition, proving that $xy \in H$ for each $x \in H$, $y \in H$, follows from the fact that $H$ is a subgroup and therefore closed under the operation of $G$.

*For the second condition, it has to be shown that e, the identity element of G, is an element of H. The proof will consist of first showing that the identity element of H is actually equal to the identity element of G.*

Let $f$ be the identity in $H$. Then $f^2 = f$. But since $f \in G$, it has an inverse $f^{-1}$ in $G$. Multiplying $f^2 = f$ on the left by $f^{-1}$, and working in $G$, $f^{-1}(ff) = f^{-1}f$ so $(f^{-1}f)f = e$, so $ef = e$, so $f = e$, and hence $e \in H$.

For the third condition, let $x \in H$. Then $x$ has an inverse, $y$, in $H$. So $xy = e$. But in $G$, there exists $x^{-1}$, the inverse of $x$ in $G$. Therefore $x^{-1} = x^{-1}e = x^{-1}(xy) = (x^{-1}x)y = ey = y$. But $y \in H$, so $x^{-1} \in H$. ∎

Here is an example showing how to use Theorem 21.

## ▆ *Example 6.3.1*

Let $a$ and $b$ be positive integers. Then $H = \{xa + yb : x, y \in \mathbf{Z}\}$ is a subgroup of $(\mathbf{Z}, +)$.

First, note that $xa + yb$ is an integer, so $H \subseteq \mathbf{Z}$.

Let $n_1, n_2 \in H$. Then $n_1 = x_1a + y_1b$ and $n_2 = x_2a + y_2b$ for integers $x_1$, $y_1$, $x_2$ and $y_2$. Therefore $n_1 + n_2 = (x_1 + x_2)a + (y_1 + y_2)b$, and since $(x_1 + x_2)$ and $(y_1 + y_2)$ are integers, $n_1 + n_2 \in H$.

Since $0 = 0a + 0b$, and $0 \in \mathbf{Z}$, $0 \in H$.

Finally, let $n = xa + yb \in H$. Consider $-n = (-x)a + (-y)b$. As $(-x)$ and $(-y)$ are integers, $(-n) \in H$. Moreover, $n + (-n) = (-n) + n = 0$, so $(-n)$ is the inverse of $n$ and $(-n) \in H$.

The conditions of Theorem 21 are satisfied so $H$ is a subgroup of $\mathbf{Z}$.

## 6.4 THE SUBGROUP GENERATED BY AN ELEMENT

An important type of subgroup is the set of all the powers of a single element.

In Section 5.7, you saw that, in the group $D_3$ the distinct powers of $S$ consisted of $S$, $S^2$ and $S^3$, where $S^3 = I$. These powers of $S$, form the group shown in the table in Fig. 6.3.

|        | $I$ | $S$ | $S^2 = R$ |
|--------|-----|-----|-----------|
| $I$    | $I$ | $S$ | $R$       |
| $S$    | $S$ | $R$ | $I$       |
| $S^2 = R$ | $R$ | $I$ | $S$    |

**Fig. 6.3**

A similar situation holds for any element in any group.

**Theorem 22**: Let $x$ be an element of a group $G$. Then $H = \left\{x^n : n \in \mathbf{Z}\right\}$ is a subgroup of $G$.

> *Informally, this says that the set of all powers of an element form a group. Notice that if the group G is finite, there are only finitely many members of H, that is, H is finite as well.*
>
> *You might think that there are two cases to consider, when the order of x is finite, and when the order is infinite, but the proof, which consists of applying Theorem 21, actually applies to both of them.*

**Proof:** Suppose that $a \in H$ and $b \in H$. Then $a = x^r$ and $b = x^s$ where $r$ and $s$ are integers, so $ab = x^{r+s}$ where, since $r$ and $s$ are integers, $r + s$ is also an integer. Therefore $ab \in H$.

As $0 \in \mathbf{Z}$, $e = x^0$ is a member of $H$.

Finally, let $a \in H$ so that $a = x^r$, where $r \in \mathbf{Z}$. Then $a^{-1} = \left(x^r\right)^{-1} = x^{-r}$. But $x^{-r} \in H$ as $-r \in \mathbf{Z}$. Therefore $a^{-1} \in H$.

As the conditions of Theorem 21 are satisfied, $H$ is a subgroup of $G$. ∎

**Definition**: The subgroup $H$ of a group $G$ defined by $H = \left\{x^n : n \in \mathbf{Z}\right\}$ is called the **subgroup generated by** $x$. It is written $\langle x \rangle$.

> *The subgroup generated by x is sometimes called the cyclic subgroup generated by x. Cyclic subgroups are discussed in Chapter 7.*

From Theorem 18 it follows that:

(1) if an element $x$ has infinite order, then $\langle x \rangle$ is infinite and $\langle x \rangle = \left\{ \ldots, x^{-2}, x^{-1}, e, x, x^2, \ldots \right\}$

(2) if an element $x$ has finite order $n$, then $\langle x \rangle = \left\{ e, x, x^2, \ldots, x^{n-1} \right\}$.

So the order of the element $x$ is equal to the order of the subgroup generated by $x$.

## ■ *Example 6.4.1*

In the group $(\mathbf{Z}, +)$, the subgroup $\langle 5 \rangle$ consists of all the multiples of 5, namely $5\mathbf{Z}$.

## ■ *Example 6.4.2*

In the group $(\mathbf{Z}_9, +)$, the subgroup $\langle 1 \rangle = \mathbf{Z}_9$, whereas $\langle 3 \rangle = \{0, 3, 6\}$.

## ■ *Example 6.4.3*

In the group $(\mathbf{Z}_5, +)$, the subgroup $\langle 3 \rangle = \mathbf{Z}_5$.

## ■ *Example 6.4.4*

In the group $D_3$ in Example 5.2.2, $\langle R \rangle = \left\{ I, R, R^2 \right\}$, and $\langle X \rangle = \left\{ I, X \right\}$.

## WHAT YOU SHOULD KNOW

■ What a subgroup is.

■ How to tell whether a subset of a group is a subgroup.

■ What is meant by the subgroup generated by an element.

## EXERCISE 6

**1** Find all the subgroups of the group shown in Fig. 5.1. Which of the subgroups is a proper subgroup?

**2** Let $G$ be an abelian group. Prove that the set of elements of order 2, together with the identity element, that is, $H = \left\{ a \in G : a^2 = e \right\}$, is a subgroup of $G$.

**3** Let $G$ be an abelian group, and let $H = \left\{ x \in G : x^3 = e \right\}$. Prove that $H$ is a subgroup of $G$.

**4** Let $G$ be an abelian group. Prove that the set of elements of finite order, $H = \left\{ a \in G : a^n = e, \text{ some } n \right\}$ is a subgroup of $G$.

**5** Mark each of the following statements true or false.

(1)   The group in Example 5.5.2 has six subgroups.

(2)   A non-abelian group can have a proper abelian subgroup.

(3)   An infinite group cannot have a finite subgroup.

(4)   An infinite proper subgroup of an infinite group must be abelian.

(5)   All groups have proper subgroups.

**6** Let $A$ and $B$ be subgroups of a group $G$. Prove that $A \cap B$ is a subgroup of $G$. Is $A \cup B$ a subgroup? Justify your answer.

**7** Let $H$ be a subgroup of a group $G$, and let $a \in H$. Prove that $\langle a \rangle \subseteq H$.

**8** Let $H$ be a subset of a group $G$. Prove that $H$ is a subgroup of $G$ if, and only if, $H$ is non-empty and $xy^{-1} \in H$ for all $x, y \in H$. (This is an alternative to Theorem 21; a more compact criterion for a subgroup.)

**9** Let $G$ be a group, and let $g$ be a fixed element of $G$. Prove that $H = \left\{ x \in G : gx = xg \right\}$ is a subgroup of $G$.

**10** Let $G$ be a group with a subgroup $K$, and let $H$ be a subgroup of $K$. Prove that $H$ is a subgroup of $G$.

# 7

# *Cyclic groups*

## 7.1 INTRODUCTION

Some symmetry groups are especially simple. Consider the Isle of Man
motif shown in Fig. 7.1.

**Fig. 7.1** *The Isle of Man motif*

If $r$ stands for a rotation about the centre of $120°$ anti-clockwise, then $e$,
$r$ and $r^2$ are symmetries of the figure as each leaves it in the same
overall position.

> *The symmetry $r^2$ is a rotation of $240°$ and $e$ is the 'do nothing'
> rotation of $0°$. Notice that $r^3 = e$.*

In fact $e$, $r$ and $r^2$ are the only distance-preserving transformations of the plane which leave this figure in the same overall position, and they form a group under the operation of 'followed by'. The symmetry group of the Isle of Man motif, $G = \{e, r, r^2\}$, where $r^3 = e$, is an example of a cyclic group; all the elements of $G$ are powers of one element in the group.

Cyclic groups are very closely related to the idea of the order of an element, which you met in Section 5.8.

## 7.2 CYCLIC GROUPS

**Definition**: Let $G$ be a group. If there is an element $g \in G$ such that every element of $G$ is of the form $g^n$ for some $n \in \mathbf{Z}$, then $G$ is called a **cyclic group**. Such an element $g$ is called a **generator** of $G$. The notation $G = \langle g \rangle$ is used to show that $g$ is a generator for $G$.

> *The elements of G consist of powers of an element of G. Notice that they could be negative as well as positive powers of g.*
>
> *To show that a group is cyclic, you have to produce a generator, and then demonstrate that it is a generator. To show that a group is not cyclic, you must show that no element is a generator.*

An element is a generator if, and only if, its order is equal to the order of the group.

### ■ *Example 7.2.1*

In the group $G = \{e, r, r^2\}$, with $r^3 = e$, there is more than one generator. You can easily check that $r^2$ is also a generator, since $(r^2)^2 = r^4 = r^3 r = er = r$, and $(r^2)^3 = r^6 = (r^3)^2 = e^2 = e$.

### ■ *Example 7.2.2*

The group $(\mathbf{Z}_7, +)$ is cyclic. You can check that the element 1 is a generator, so $\mathbf{Z}_7 = \langle 1 \rangle$.

In general the group $(\mathbf{Z}_n, +)$ is a cyclic group. Once again, 1 is a generator, so $\mathbf{Z}_n = \langle 1 \rangle$.

## ■ *Example 7.2.3*

A cyclic group can be infinite. An example is $(\mathbf{Z}, +)$. The number 1 is a generator, because every element $n \in \mathbf{Z}$ can be written in the form $n = n(1)$. Therefore $(\mathbf{Z}, +) = \langle 1 \rangle$.

## ■ *Example 7.2.4*

The group consisting of the six 6th roots of unity, $\left\{ z \in \mathbf{C} : z^6 = 1 \right\}$, is a cyclic group. The elements of this group are $1, w, w^2, w^3, w^4, w^5$ where $w = e^{i\pi/3}$. So $w$ is a generator. The element $w^5$ is also a generator, because every element of the group is also a power of $w^5$. But $w^2$ is not a generator, because its distinct powers are $\left( w^2 \right)^1 = w^2$, $\left( w^2 \right)^2 = w^4$ and $\left( w^2 \right)^3 = 1$. And $w^3$ is not a generator, because its distinct powers are $\left( w^3 \right)^1 = w^3$ and $\left( w^3 \right)^2 = 1$.

## ■ *Example 7.2.5*

The group $D_3$ in Example 5.2.2 is not cyclic. For, using the notation of Fig. 5.3, neither $R$ nor $S$ is a generator, since they both have order 3, and none of $X$, $Y$ and $Z$ is a generator, since each has order 2.

## ■ *Example 7.2.6*

The group $(\mathbf{Q}^*, \times)$ is not cyclic. For suppose $p/q$ with $p$ and $q$ relatively prime is a generator for $\mathbf{Q}^*$. Then every element of $\mathbf{Q}^*$ can be expressed as a power of $p/q$. Either $|p/q| < 1$ or $|p/q| \geq 1$. If $|p/q| < 1$, all powers of $p/q$ have magnitude less than 1, and no power of $p/q$ will be equal to 2. Similarly, if $|p/q| \geq 1$, all powers of $p/q$ have magnitude greater than or equal to 1, and no power of $p/q$ will be equal to $1/2$. Therefore there is no generator, and $(\mathbf{Q}^*, \times)$ is not cyclic.

## 7.3 SOME DEFINITIONS AND THEOREMS ABOUT CYCLIC GROUPS

**Theorem 23**: Every cyclic group is abelian.

**Proof**: Let $G$ be a cyclic group, with generator $g$, and let $x, y \in G$. Then $x = g^m$ and $y = g^n$ for some integers $m$ and $n$. Then $xy = g^m g^n = g^{m+n} = g^n g^m = yx$, so, since $xy = yx$ for all $x, y \in G$, $G$ is abelian. ∎

**Theorem 24**: Every subgroup of a cyclic group is cyclic.

**Proof**: Suppose that $G$ is a cyclic group, and that $H$ is a subgroup of $G$. Suppose that $g$ is a generator of $G$.

If $H$ consists of the identity element $e$ alone, then $H = <e>$, and $H$ is cyclic. Suppose that $H$ contains an element other than the identity element.

> *Remember that to show that H is cyclic, you have to produce a generator. As it seems likely that the least power of g which belongs to H will be the generator, that is the power to look for. You can therefore expect to use in the process the axiom about a set of positive integers having a least member.*

Then $g^n \in H$ for some $n \in \mathbf{Z}$, $n \neq 0$. You can suppose that $n > 0$, since both $g^n$ and $g^{-n} \in H$. Let $m$ be the smallest positive integer for which $g^m \in H$. We claim that $g^m = a$, say, is a generator of $H$.

> *So you now have to prove that H = <a>, that is, that two sets are equal. You have therefore to prove that each is a subset of the other. From the way that a was defined, it is clear that $<a> \subseteq H$. See also question 7 in Exercise 6. It therefore remains to show that $H \subseteq <a>$, that is, that every element of H is a power of a.*

Suppose that $x \in H$. Then, since $H$ is a subgroup of $G$, $x = g^N$ for some $N$. From the division algorithm, you can write $N = qm + r$, where $0 \leq r < m$. As $g^m \in H$, $\left(g^m\right)^q \in H$, and hence $\left(\left(g^m\right)^q\right)^{-1} \in H$. But $g^N \in H$ and $g^r = g^{N-qm} = g^N g^{-qm} = g^N \left(\left(g^m\right)^q\right)^{-1}$.

Therefore $g^r \in H$. But $0 \le r < m$ and $m$ is the least positive integer for which $g^m \in H$. Therefore $r = 0$, and $x = g^N = g^{qm+r} = \left(g^m\right)^q = a^q$, so $x$ can be written as a power of $a$. ∎

Here is an application of this theorem.

## ■ *Example 7.3.1*

In Example 6.3.1, it was shown that, if $a$ and $b$ are positive integers, then $H = \{xa + yb : x, y \in \mathbf{Z}\}$ is a subgroup of $(\mathbf{Z}, +)$.

Using Theorem 24, $H$ is a cyclic group, and so has a generator $h$, which can be assumed to be positive. Therefore $\{xa + yb : x, y \in \mathbf{Z}\} = h\mathbf{Z}$.

The generator $h$ is actually the highest common factor of $a$ and $b$, for the following reasons. Since $a \in \{xa + yb : x, y \in \mathbf{Z}\}$ by taking $x = 1$ and $y = 0$, it follows that $a \in h\mathbf{Z}$ because $\{xa + yb : x, y \in \mathbf{Z}\} = h\mathbf{Z}$. Therefore $h$ divides $a$. Similarly $h$ divides $b$. But is it the highest common factor? Suppose that $d$ is another common factor of $a$ and $b$. Then $d$ divides all numbers of the form $xa + yb$. But $h$ is of the form $xa + yb$ because $h \in h\mathbf{Z}$ and $h\mathbf{Z} = \{xa + yb : x, y \in \mathbf{Z}\}$. Therefore $d$ divides $h$. So $d$ is less than or equal to $h$, and $h$ is therefore the highest common factor.

The fact that $h$ is an element of $H$ means that the highest common factor of $a$ and $b$ can be written as an integer linear combination of $a$ and $b$. Now $1 \in h\mathbf{Z}$ if, and only if, $h = 1$. Also, by definition, two positive integers are relatively prime if, and only if, their highest common factor is 1. So you can deduce, as a special case of Example 7.3.1, the result of Theorem 1 that $a$ and $b$ are relatively prime if, and only if, there is an integer linear combination of $a$ and $b$ which is equal to 1.

## WHAT YOU SHOULD KNOW

- ■ What a cyclic group is.

- ■ The meaning of 'generator'.

■ How to prove that a group is cyclic.

■ How to prove that a group is not cyclic.

## EXERCISE 7

**1**  Prove that the group $\left(\{2,4,6,8\}, \times \bmod 10\right)$, in Fig. 5.1, is cyclic.

**2**  Find four proper subgroups of a cyclic group of order 12.

**3**  Which elements of the cyclic group $\mathbf{Z}_6$ are generators?

**4**  Mark each of the following statements true or false.

(1)  Every cyclic group has a generator.

(2)  Every member of every cyclic group is a generator.

(3)  A cyclic group can have more than one generator.

(4)  A cyclic group can have a non-cyclic subgroup.

(5)  Every cyclic group is abelian.

(6)  $(\mathbf{Z}, +)$ is not cyclic.

(7)  $(\mathbf{C}^*, \times)$ is cyclic.

**5**  Prove that $(\mathbf{R}, +)$ is not a cyclic group.

**6**  Which of the following groups are cyclic? Give reasons.

(1)  The group $\{1, 5, 7, 11\}$, under multiplication mod 12.

(2)  The additive group $\mathbf{Q}$ of rational numbers.

(3)  The circle group $T = \left\{z \in \mathbf{C} : |z| = 1\right\}$ under multiplication of complex numbers.

(4)  $\mathbf{Z}[i] = \{a + bi : a, b \in \mathbf{Z}\}$ under addition of complex numbers.

# 8

# *Products of groups*

## 8.1 INTRODUCTION

You will be familiar with the idea of coordinates in the plane consisting of ordered pairs of numbers. You are used to plotting points such as $(3,2)$ and $\left(\sqrt{2}, -\sqrt{2}\right)$ on coordinate axes on graph paper. Each of the numbers in such a coordinate pair comes from a set: when you plot graphs this set is usually the real numbers $\mathbf{R}$. The set of all pairs of coordinates such as $(3,2)$ and $\left(\sqrt{2}, -\sqrt{2}\right)$ is called $\mathbf{R} \times \mathbf{R}$, meaning that the first number comes from the set $\mathbf{R}$, as does the second number. The notation $\mathbf{R}^2$ is sometimes used for $\mathbf{R} \times \mathbf{R}$. The Cartesian product generalises this idea.

## 8.2 THE CARTESIAN PRODUCT

The **Cartesian product** of two sets $A$ and $B$, written $A \times B$ is the set defined by

$$A \times B = \left\{(a,b) : a \in A, b \in B\right\}.$$

## ■ *Example 8.2.1*

Suppose that $A = \{2,3,4\}$ and $B = \{x, y\}$. Then:
$$A \times B = \{(2, x), (2, y), (3, x), (3, y), (4, x), (4, y)\}.$$

Notice that the set $B \times A$ is not the same as $A \times B$, because $B \times A = \{(x, 2), (y, 2), (x, 3), (y, 3), (x, 4), (y, 4)\}$. So the order of the sets in a Cartesian products matters.

Notice also that neither $A$ nor $B$ is a subset of $A \times B$. The set $A \times B$ consists of pairs: the elements of $A$ and $B$ are not pairs, so do not belong to $A \times B$.

## ■ *Example 8.2.2*

In the introduction to this section you saw the set $\mathbf{R} \times \mathbf{R}$. In the same way the set $\mathbf{Z}_2 \times \mathbf{Z}_2$ is the Cartesian product of $\mathbf{Z}_2$ with itself. So $\mathbf{Z}_2 \times \mathbf{Z}_2 = \{(0,0), (0,1), (1,0), (1,1)\}$.

## ■ *Example 8.2.3*

In the case when the sets $A$ and $B$ are finite, you can draw a conclusion about the number of elements in the set $A \times B$. For if $A$ has $m$ elements and $B$ has $n$ elements, the set $A \times B$ has $mn$ elements.

You can also extend the definition of Cartesian product in an obvious way to produce a Cartesian product of more than two sets.

## 8.3 DIRECT PRODUCT GROUPS

Suppose now that the sets in a Cartesian product are both groups, say $G$ and $H$. Is the product $G \times H$ a group?

Suppose that the two groups are $(\mathbf{Z}_3, +)$ and $(\mathbf{Z}_2, +)$. Then the elements of $\mathbf{Z}_3 \times \mathbf{Z}_2$ are $(0,0)$, $(1,0)$, $(2,0)$, $(0,1)$, $(1,1)$, $(2,1)$. But what can you suggest for the rule of composition?

Suppose that you want to find the product of $(1,1) \circ (0,2)$. You can combine the first elements in $\mathbf{Z}_2$ to get 1, and the second pair in $\mathbf{Z}_3$ to get 0, thus giving a suggested product of $(1,0)$.

If you write out a table, what you get is shown in Fig. 8.1.

You can easily check that this is a group table. The group $\mathbf{Z}_3 \times \mathbf{Z}_2$ has order 6.

|       | (0,0) | (1,0) | (2,0) | (0,1) | (1,1) | (2,1) |
|-------|-------|-------|-------|-------|-------|-------|
| (0,0) | (0,0) | (1,0) | (2,0) | (0,1) | (1,1) | (2,1) |
| (1,0) | (1,0) | (2,0) | (0,0) | (1,1) | (2,1) | (0,1) |
| (2,0) | (2,0) | (0,0) | (1,0) | (2,1) | (0,1) | (1,1) |
| (0,1) | (0,1) | (0,1) | (2,1) | (0,0) | (1,0) | (2,0) |
| (1,1) | (1,1) | (2,1) | (0,1) | (1,0) | (2,0) | (0,0) |
| (2,1) | (2,1) | (0,1) | (1,1) | (2,0) | (0,0) | (1,0) |

**Fig. 8.1**

This leads to a more general argument. Combine $(g_1, h_1)$ and $(g_2, h_2)$ by saying $(g_1, h_1) \circ (g_2, h_2) = (g_1 g_2, h_1 h_2)$ where the first multiplication is carried out in $G$ and the second is carried out in $H$; this can be described by saying that 'multiplication is carried out in components'.

**Theorem 25**: Let $G$ and $H$ be groups. Then the set $G \times H$ with the operation $(g_1, h_1) \circ (g_2, h_2) = (g_1 g_2, h_1 h_2)$ is a group. This group is called the **direct product** of the groups $G$ and $H$.

**Proof**: The first question is, 'Is the group operation well defined?' The answer is yes: as $G$ and $H$ are groups, the elements $g_1 g_2$ and $h_1 h_2$ are elements of $G$ and $H$ respectively, so $(g_1 g_2, h_1 h_2)$ belongs to $G \times H$.

Multiplication by components is an associative operation.
$$\big((g_1, h_1) \circ (g_2, h_2)\big) \circ (g_3, h_3) = (g_1 g_2, h_1 h_2) \circ (g_3, h_3) = \big((g_1 g_2)g_3, (h_1 h_2)h_3\big)$$
$$(g_1, h_1) \circ \big((g_2, h_2) \circ (g_3, h_3)\big) = (g_1, h_1) \circ (g_2 g_3, h_2 h_3) = \big(g_1(g_2 g_3), h_1(h_2 h_3)\big).$$
These last two expressions are equal, because the operations within $G$ and $H$ are themselves associative.

The identity element is $(e_G, e_H)$, because, for any $(g, h) \in G \times H$, $(e_G, e_H) \circ (g, h) = (e_G g, e_H h) = (g, h)$. Also, $(g, h) \circ (e_G, e_H) = (g, h)$. Therefore $(e_G, e_H)$ is the identity for $G \times H$.

Finally, the element $\left(g^{-1}, h^{-1}\right) \in G \times H$ is the inverse of $(g, h)$. For $\left(g^{-1}, h^{-1}\right) \circ (g, h) = \left(g^{-1}g, h^{-1}h\right) = \left(e_G, e_H\right)$. Similarly $(g, h) \circ \left(g^{-1}, h^{-1}\right)$ $= \left(e_G, e_H\right)$, so $\left(g^{-1}, h^{-1}\right)$ is the inverse of $(g, h)$ in $G \times H$.

So $G \times H$ with the operation of multiplication by components, is a group. ∎

Direct products allow you to build bigger groups out of smaller ones. You will see in Chapter 11 that they also sometimes enable you to describe the structure of a complicated group in terms of the structures of more familiar smaller groups.

## WHAT YOU SHOULD KNOW

- The meaning of and the notation for Cartesian product.

- How to carry out operations in a direct product of groups.

## EXERCISE 8

**1** Mark each of the following statements true or false.

(1) The set $A \times B$ always has a finite number of elements.

(2) The set $\mathbf{Z}_m \times \mathbf{Z}_n$ has $mn$ elements.

(3) If the group $G$ has $m$ elements, and the group $H$ has $n$ elements, then $G \times H$ has $mn$ elements.

(4) You cannot form the direct product $G \times H$ if $G$ is infinite.

**2** Write a definition for the set $A \times B \times C$. Use your definition to write out the elements of $\mathbf{Z}_2 \times \mathbf{Z}_2 \times \mathbf{Z}_2$.

**3** Prove that $\mathbf{Z} \times \mathbf{Z} \subseteq \mathbf{Q} \times \mathbf{Q}$.

**4** You can interpret the set $\mathbf{R} \times \mathbf{R}$ as all the points on an infinite sheet of paper. How would you interpret $\mathbf{Z} \times \mathbf{Z}$?

**5** Write out a group table for $\mathbf{Z}_2 \times \mathbf{Z}_2$.

**6** Prove that if $G$ and $H$ are abelian, then $G \times H$ is also abelian.

**7** Prove that if $G$ is not abelian, then $G \times H$ is not abelian.

**8** Write out a group table for $\mathbf{Z}_2 \times \mathbf{Z}_3$. Is it the same group as $\mathbf{Z}_3 \times \mathbf{Z}_2$?

**9** Show that $(1,1)$ is a generator for $\mathbf{Z}_3 \times \mathbf{Z}_2$. Are there any other generators?

**10** Prove that if $(x, y) \in G \times H$ has order $n$ in $G \times H$, then the order of $x \in G$ divides $n$, and the order of $h \in H$ divides $n$.

# 9

# *Functions*

## 9.1 INTRODUCTION

It is likely that you already know a definition of function, probably in the context of functions of real numbers and their graphs. If that is the case, you will see that the definition given in this chapter is a generalisation. If, on the other hand, you have already met the generalised idea of function, then this chapter is likely to be a revision of ideas, notation and language.

In some contexts functions are called mappings. However, the words 'function' and 'mapping' have the same meaning.

## 9.2 FUNCTIONS: A DISCUSSION

In the context of graphs, people talk about a function of $x$ such as $f(x) = x^2$. But what are the properties that $f$ must have in order to be called a function?

First, what is $x$? In the case $f(x) = x^2$ there is an understanding that $x$ is a real number, but in the case $g(x) = \sqrt{x}$, the positive square root of $x$, there is an understanding that $x$ is a positive real number or zero. The point is that in each case $x$ is a member of a starting or object set, although you may not be told explicitly what this starting or object set is. In the generalisation which follows, $x$ will be taken from an object set which will be called the **domain**, and you will be told explicitly what the domain is.

Secondly, what about the result $x^2$ obtained by operating on $x$ with the function $f$? It also belongs to a set: this is sometimes called the target set, but more often the **co-domain**. Once again, you should be told explicitly what the co-domain is. In the example $f(x) = x^2$ the co-domain may be **R**, or it may be $\mathbf{R}^+ \cup \{0\}$. In the context of calculus, you may not be told what the domain is, and often it will not matter precisely what the co-domain is; in the context of abstract algebra you will always be told the target set or co-domain.

Thirdly, when you are told that $f$ is a function, you expect that for every value of $x$ you have a rule for calculating the value $f(x)$ of the function corresponding to $x$. There are two points of emphasis to be made here. The rule must enable you to calculate the value of $f(x)$ for **every** value of $x$. And you must be able to calculate **the** value of $f(x)$ for every value of $x$. Both these aspects are important in the generalised definition of a function. There is a rule which for each value of $x$ in the domain gives you just one value of $f(x)$ in the co-domain. Thus $f(x)$ is called the **image** of $x$ under $f$.

Fig. 9.1 illustrates these three aspects of the idea of a function. You can see:

● the starting set, the domain $A$
● the target set, the co-domain $B$
● the rule showing the element $b \in B$ corresponding to $a \in A$.

The fact that for each element $x$ in the domain there corresponds just one value of $f(x)$ in the co-domain means that there is exactly one arrow leaving every point in the domain $A$, with its end in the co-domain $B$.

Notice that it is not necessary for every element in $B$ to be the image of an element of $A$.

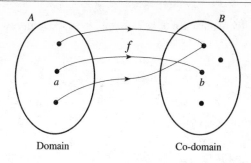

**Fig. 9.1** *Illustration of a function*

## 9.3 FUNCTIONS: A DEFINITION

**Definition**: A **function** $f$ has three components: a starting set $A$, called the **domain**; a target set $B$, called the **co-domain**; and a **rule** which assigns to each member $a$ of $A$ a unique member $b$ of $B$.

> *Notice that it is hard to say precisely what a 'rule' is. A rule is to be understood as no more than an association of elements of B with elements of A for which (1) every element of A has an associated element of B, and (2) no element of A has more than one associated element of B.*

## 9.4 NOTATION AND LANGUAGE

You write $f : A \to B$ to show that $f$ is a function for which the domain is $A$ and the co-domain is $B$.

You say that $f$ maps $a$ to $b$, and that $f$ maps $A$ into $B$. The notation to indicate that $a$ maps to $b$ is $b = f(a)$.

The set
$$\{b \in B : b = f(a) \text{ for some } a \in A\}$$
is called the **set of images** or **image set** of $A$. It is also called the **range** of $f$.

Two functions $f : A \to B$ and $g : C \to D$ are **equal** if $A = C$, $B = D$ and $f(x) = g(x)$ for every element $x$ in $A$.

In Fig. 9.1 the image set of $A$ is a subset of $B$ containing just two elements.

A natural extension of the idea of image set that you will meet later in the book is that of the **image of a subset** of $A$. Let $f : A \to B$, and let $X \subseteq A$. Then the image $f(X)$ of the subset $X$ is defined by $f(X) = \{f(x) : x \in X\}$.

**Definition**: The function $I_A : A \to A$ given by $I_A(x) = x$ for all $x \in A$ is called the **identity function** on $A$.

*If there is no ambiguity, the subscript A can be omitted.*

## 9.5 EXAMPLES

### ■ *Example 9.5.1*

In Section 9.2 in the discussion of the two examples $f(x) = x^2$ and $g(x) = \sqrt{x}$, complete definitions of $f$ and $g$ might be as follows:

$f : \mathbf{R} \to \mathbf{R}$ such that $f(x) = x^2$

$g : \mathbf{R}^+ \cup \{0\} \to \mathbf{R}$ such that $g(x) = \sqrt{x}$, the positive square root of $x$.

Notice that for both the functions $f$ and $g$, the set of images is not the same as the whole domain $\mathbf{R}$. You could have defined the function $f$ differently by saying that:

$f : \mathbf{R} \to \mathbf{R}^+ \cup \{0\}$ such that $f(x) = x^2$

but in that case, the two functions called $f$ are different. This illustrates an important fact: that to be equal, two functions must have the same domain, the same co-domain and the same rule.

### ■ *Example 9.5.2*

Consider the function $f : \mathbf{Z} \to \mathbf{Z}_n$ such that $f(x) = [x]_n$, the residue class of $x$ modulo $n$.

$\mathbf{Z}_n$ consists of the $n$ residue classes $\{[r] : r = 0, 1, 2, \dots, n-1\}$. Each of these residue classes is in the range of $f$ because, for each $r$, $f(r) = [r]$.

In fact for each element $[r] \in \mathbf{Z}_n$ there are infinitely many elements of $\mathbf{Z}$ which map to it because $f(x) = [r]$ for every element $x$ in the set $r + n\mathbf{Z}$, that is, for every integer of the form $x = r + kn$.

Here are some examples of 'functions' which are not well defined.

## ■ *Example 9.5.3*

Let $f : \mathbf{Q} \to \mathbf{Z}$ be defined by $f(x) =$ the numerator of the fraction $x$.

This example fails the uniqueness part of the definition of a function.

## ■ *Example 9.5.4*

Let $f : \mathbf{Z} \to \mathbf{Q}$ be given by $f(n) = 1/n$.

This is not defined for $n = 0$.

## ■ *Example 9.5.5*

Let $X =$ set of times on a specified day, and $Y =$ set of trains leaving Victoria Station in London on that day. Then let $f : X \to Y$ be given by $f(x) =$ train leaving Victoria Station at time $x$.

This example fails on two grounds: given a particular time, there may not be a train leaving Victoria Station at that time; and there may be some times at which more than one train leaves Victoria Station.

## 9.6 INJECTIONS AND SURJECTIONS

**Definition**: A function $f : A \to B$ is called an **injection** if each element of $B$ has at most one element of $A$ mapped into it. The adjective **injective** is used to describe a function which is an injection.

Fig. 9.2 illustrates an injection. Each element of $B$ has at most one arrow coming into it from an element of $A$. Notice that there may be elements of $B$ which are not images of elements of $A$.

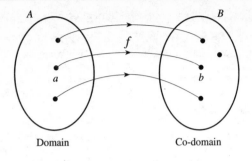

**Fig. 9.2** *The function* $f : A \to B$ *is an injection*

Proving that a given function *f* is an injection is equivalent to proving that if $f(a) = f(b)$ then $a = b$. Simply writing down $f(a) = f(b)$ as the first line is often a good way to start the proof. Example 9.6.1 illustrates this.

**Definition**: A function $f : A \to B$ is called a **surjection** if each element of *B* has at least one member of *A* mapped into it. The adjective **surjective** is used to describe a function which is an surjection.

Fig. 9.3 illustrates a surjection. Each element of *B* has at least one arrow coming into it from an element in *A*. Some elements in *B* may be the image of more than one element in *A*.

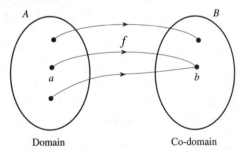

**Fig. 9.3** *The function* $f : A \to B$ *is a surjection*

**Definition**: A function $f : A \to B$ which is both an injection and a surjection is called a **bijection**. The adjective **bijective** is used to describe a function which is a bijection.

Fig. 9.4 illustrates a bijection. Each element of *B* has exactly one arrow coming into it from an element of *A*.

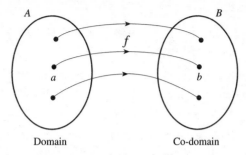

**Fig. 9.4** *The function* $f : A \to B$ *is a bijection*

> *Sometimes the term 'onto function' is used instead of surjection, and the term 'one to one correspondence' is used to mean bijection. In this book, the terms 'surjection' and 'bijection' will be used.*

## ■ *Example 9.6.1*

The function $f : \mathbf{Z} \to \mathbf{Z}$ such that $f(n) = n + 1$ is both an injection and a surjection.

To prove that *f* is an injection, suppose that *a* and *b* map to the same image under *f*. Then $f(a) = f(b)$ so $a + 1 = b + 1$, leading to $a = b$.

> *This method is one you will generally use for proving that a function is an injection. You should start by supposing that two different elements have identical images, and then show that the elements themselves must be identical.*

To prove that *f* is a surjection, you must find the element which maps onto any given element in the co-domain. So if you take an element *n* in the co-domain, consider the element $n - 1$ in the domain. Then $f(n - 1) = (n - 1) + 1 = n$.

> *This method is fairly typical for proving that a function f is a surjection. For any given element of the co-domain, you have to identify an element in the domain which maps onto it.*

In this case, as $f$ is both an injection and a surjection, $f$ is a bijection.

## ■ *Example 9.6.2*

The projection function $p : \mathbf{R}^2 \to \mathbf{R}^2$ defined by $p(x, y) = (x, 0)$ is neither an injection nor a surjection.

To prove that $p$ is not an injection, you need to find two elements in the domain which have the same image under $p$. Since $p(0, 1) = (0, 0)$ and $p(0, 0) = (0, 0)$, $p$ is not an injection.

To prove that $p$ is not a surjection, you need to find an element in the co-domain $\mathbf{R}^2$ which is not the image of any element of the domain $\mathbf{R}^2$. Consider the element $(0, 1)$ and suppose it is the image of $(x, y)$. Then $p(x, y) = (x, 0) = (0, 1)$, which implies that $x = 0$ and $0 = 1$. As the second of these is impossible, no such element exists, and $p$ is not a surjection.

## ■ *Example 9.6.3*

Let $G$ be a group, and let $g$ be any element of $G$. Prove that the function $f : G \to G$ defined by $f(x) = gx$ is a bijection.

*Injection.* Suppose that $f(x) = f(y)$. Then $gx = gy$, and, by Theorem 15, part (4), $x = y$. Therefore $f$ is injective.

*Surjection.* Suppose that $y$ is any member of $G$. Then $g^{-1}y \in G$, and $f(g^{-1}y) = g(g^{-1}y) = ey = y$, so f is surjective.

Therefore, as $f$ is both injective and surjective, $f$ is a bijection.

*When $G$ is finite, the elements of the form $gx$ in the group table for $G$ are in the row with $g$ at the left-hand end. The fact that $f$ is a bijection tells you that all the elements are in the row, and that no element is repeated.*

*You can prove the same property for columns by showing that $f : G \to G$ defined by $f(x) = xg$ is also a bijection.*

## 9.7 INJECTIONS AND SURJECTIONS OF FINITE SETS

Here are two theorems about surjective and injective functions when the domain and co-domain are finite sets. You may find it useful to draw diagrams like Figs. 9.2 to 9.4 to illustrate the theorems for yourself. The results may then appear to be obvious.

Before the theorems, here is a piece of notation. Let $X$ be a finite set. Then $|X|$ means the number of elements in $X$.

**Theorem 26**: Let $X$ and $Y$ be finite sets, and let $f : X \to Y$ be a function. Then:

(1) $|f(X)| \leq |X|$

(2) $f$ is injective if, and only if, $|f(X)| = |X|$

(3) $f$ is surjective if, and only if, $|f(X)| = |Y|$.

**Proof**: (1) By the definition of function, to each $x \in X$ there is assigned one, and only one, element of $Y$. So to each element of $f(X)$, there corresponds at least one element of $X$; and no two different elements of $f(X)$ are associated with the same element of $X$. Therefore $|f(X)| \leq |X|$.

(2) *If*. Suppose that $|f(X)| = |X|$. Then $X$ and $f(X)$ have the same number of elements. Suppose that $f(x_1) = f(x_2)$ for $x_1, x_2 \in X$. If $x_1 \neq x_2$, then $|f(X)| \leq |X|$, since the images of two different elements in $X$ would be the same. But as $|f(X)| = |X|$, $x_1 = x_2$, and $f$ is an injection.

*Only if*. Suppose that $f$ is injective. Then, if $f(x_1) = f(x_2)$, $x_1 = x_2$. Therefore the elements of $f(X)$ are all different, and $|f(X)| = |X|$.

(3) *If*. Suppose that $|f(X)| = |Y|$. Then $f(X)$ and $Y$ have the same number of elements. But $f(X) \subseteq Y$. Therefore $f(X) = Y$. Therefore every element of $Y$ is the image of some member $x \in X$. Therefore $f$ is a surjection.

*Only if.* Suppose that $f$ is surjective. Then every element of $Y$ is the image of some element of $X$. So $Y \subseteq f(X)$. But $f(X) \subseteq Y$. Therefore $f(X) = Y$, so $|f(X)| = |Y|$. ∎

Since, by definition, $f(X) \subseteq Y$, and so $|f(X)| \leq |Y|$, you can deduce that if $f$ is an injection, $|X| \leq |Y|$. Or, put an equivalent way, if $|X| > |Y|$, then $f$ is not an injection.

This result is called the **Pigeon-hole principle**, and is stated in the following way, with the justification in brackets.

Suppose that you have pigeons which are put into pigeon-holes (that is, a function $f : X \to Y$), and you have more pigeons than pigeon-holes (that is, $|X| > |Y|$), then there is at least one pigeon-hole with more than one pigeon (that is, $f$ is not injective).

The Pigeon-hole principle has some quite surprising applications. Here is a none-too-serious example.

## ■ *Example 9.7.1*

Let $X$ = the set of non-bald people in England, and let $Y$ = the set of positive integers less than a million. Let $f : X \to Y$ be given by $f(x)$ = the number of hairs on the head of $x$.

It is a fact that no-one has more than a million hairs on his or her head, (the number is usually 150 000 to 200 000), so $f$ is well defined.

It is also a fact that $|X| > |Y|$, that is, the number of non-bald people in England is greater than a million.

It follows from the Pigeon-hole principle that $f$ is not injective, and hence that there are two people in England with the same number of hairs on their heads.

**Theorem 27**: Let $X$ and $Y$ be finite sets, let $|X| = |Y|$, and let $f : X \to Y$ be a function. Then $f$ is injective if, and only if, $f$ is surjective.

**Proof**: *If.* Suppose that $f$ is surjective. Then, by Theorem 26, part (3), $|f(X)| = |Y|$. But, by hypothesis, $|X| = |Y|$, so $|f(X)| = |X|$. Therefore, by Theorem 26, part (2), $f$ is injective.

*Only if.* Suppose that $f$ is injective. Then, by Theorem 26, part (2), $|f(X)| = |X|$. But, by hypothesis, $|X| = |Y|$, so $|f(X)| = |Y|$. Then, by Theorem 26, part (3), $f$ is surjective. ∎

## WHAT YOU SHOULD KNOW

- The meaning of the word 'function' or 'mapping'.

- The language and notation associated with functions.

- The meanings of the terms 'injection', 'surjection' and 'bijection'.

- How to prove whether a given function is an injection, a surjection, a bijection, or none of these.

- What the 'Pigeon-hole principle' is.

## EXERCISE 9

**1**  Prove that $f : \mathbf{R} \to \mathbf{R}$ defined by $f(x) = \sin x$ is neither an injection nor a surjection.

**2**  Give an example of a function $f : \mathbf{R} \to \mathbf{R}$ which is an injection, but not a surjection.

**3**  Give an example of a function $f : \mathbf{R} \to \mathbf{R}$ which is a surjection, but not an injection.

**4**  In each of the following examples the domain is $\mathbf{R}$ and the co-domain is $\mathbf{R}$. Decide which of them is the definition of a function. If it is a function, decide whether it is injective, and whether it is surjective. Be able to justify your answers. Keep your answers: they will be used in question 2 of Exercise 10.

(1)  $f(x) = x^2$  (2)  $f(x) = x^3$

(3)  $f(x) = 1/x$  (4)  $f(x) = \cos x$

(5)  $f(x) = \tan x$  (6)  $f(x) = e^x$

(7)  $f(x) = |x|$

(8)  $f(x) = \operatorname{int} x$  ( $\operatorname{int} x$ is the largest integer $\leq x$.)

(9)  $f(x) = \sqrt{x}$  (10)  $f(x) = x + 1$

(11)  $f(x) = \sin^{-1} x$

(12)  $f(x)$ is the smallest real number greater than $x$.

**5** Prove that the function $f : \mathbf{R}^+ \cup \{0\} \to \mathbf{R}^+ \cup \{0\}$ defined by $f(x) = \sqrt{x}$ is a bijection.

**6** Prove that the function $f : \mathbf{R}^* \to \mathbf{R}^*$ defined by $f(x) = 1/x$ is a bijection.

**7** Prove that the function $f : \mathbf{Z} \to \mathbf{Z}^+$ defined by

$$f(n) = \begin{cases} 2n, & \text{if } n > 0 \\ 1 - 2n, & \text{if } n \le 0 \end{cases}$$

is a bijection.

**8** How can you tell from the graph of a function $y = f(x)$, whether or not $f$ is (1) an injection; (2) a surjection; (3) a bijection?

**9** Mark each of the following statements true or false.

(1) The function $f : \mathbf{R} \to \mathbf{R}$ such that $f(x) = 0$ for all $x \in \mathbf{R}$, is a bijection.

(2) The function $f : \mathbf{Z}_3 \to \mathbf{Z}_3$ such that $f(x) = x + 1$, is an injection.

(3) Every function which is a bijection is also a surjection.

(4) $f : \mathbf{R} \times \mathbf{R} \to \mathbf{R} \times \mathbf{R}$ defined by $f(x, y) = (y, x)$, is a bijection.

**10** Let $P = \{\text{polynomials in } x \text{ with real coefficients}\}$. Decide, in each of the following cases, whether $f : P \to P$ defined in the following ways is a function. If it is, decide whether it is injective, and whether it is surjective.

(1) $f(p) = \dfrac{d(p(x))}{dx}$

(2) $f(p) = \displaystyle\int p(x)\, dx$

(3) $f(p) = \displaystyle\int_0^x p(t)\, dt$

(4) $f(p) = x p(x)$

**11** Let $G$ be a group, and let $g$ be any element of $G$. Prove that the function $f : G \to G$ defined by $f(x) = xg$ is a bijection.

# 10

# *Composition of functions*

## 10.1 INTRODUCTION

When the co-domain of one function is the domain of another function you can combine the two functions in the way indicated by Fig. 10.1.

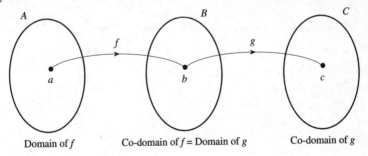

**Fig. 10.1**

In Fig. 10.1 $b = f(a)$ and $c = g(b)$. The composite function is the function which maps $a$ from the set $A$ directly to $c$ in set $C$.

These ideas are set up more formally in the next section.

## 10.2 COMPOSITE FUNCTIONS

First you have to establish that the rule which takes $a$ to $c$ is actually a properly defined function.

**Theorem   28** Let $f : A \to B$ and $g : B \to C$ be functions. Then $h : A \to C$ such that $h(x) = g\{f(x)\}$ is a function.

**Proof**: To show that $h$ is a function, you need to exhibit its domain and its co-domain, and to show that for every element in the domain there is a unique element in the co-domain. In this case the domain is $A$, and the co-domain is $C$, so it only remains to show the existence and uniqueness properties. For each element $a \in A$ there exists a unique element $f(a) = b$ in $B$. Moreover, for that $b \in B$ there is a unique element $g(b) = c$ in $C$. It follows that for each element $x \in A$ there exists a unique element $c \in C$, such that $c = g\{f(a)\}$.

This shows that $h : A \to C$ such that $h(x) = g\{f(x)\}$ is a function. ∎

Let $f : A \to B$ and $g : B \to C$ be functions. Then the function $h : A \to C$ such that $h(x) = g\{f(x)\}$ is called the **composite** of $f$ and $g$, and written sometimes as $g \circ f$ and sometimes as $gf$. Then $(g \circ f)(x) = g\{f(x)\}$.

> *The composite function $g \circ f$ is pronounced 'g circle f'. In this chapter and the next the notation $g \circ f$ will be used for clarity. Subsequently the notation $gf$ will be used for brevity.*

Notice that for the composite function to be properly defined it is not necessary that the function $f : A \to B$ is a surjection. For example, let $f : \mathbf{R} \to \mathbf{R}$ be given by $f(x) = \sin x$ and $g : \mathbf{R} \to \mathbf{R}$ be given by $g(x) = 2x$. Then the function $(g \circ f) : \mathbf{R} \to \mathbf{R}$ is defined by $(g \circ f)(x) = g\{f(x)\} = 2\sin x$, even though the image set of $f$ is the set $\{x \in \mathbf{R} : -1 \le x \le 1\}$.

In the case of functions $f : A \to A$ and $g : A \to A$, $gf$ and $fg$ are both functions $A \to A$, but they need not be equal. For example, consider $f : \mathbf{R} \to \mathbf{R}$ where $f(x) = 2x + 1$ and $g : \mathbf{R} \to \mathbf{R}$ given by $g(x) = 2x$. Then $(fg)(1) = f(g(1)) = f(2) = 5$ and $(gf)(1) = g(f(1)) = g(3) = 6$.

## 10.3 SOME PROPERTIES OF COMPOSITE FUNCTIONS

What can you deduce about the composite function $g \circ f : A \to C$ when you know some of the properties of the functions $f : A \to B$ and $g : B \to C$? For example, if $f$ and $g$ are both surjections, will $g \circ f$ be a surjection?

*The answer is yes. The result itself is not very important, but the method of proof is important. As in all proofs about surjections, you have to find an element in A which maps to a given element in the image C.*

### ■ *Example 10.3.1*

Prove that if $f : A \to B$ and $g : B \to C$ are surjections, then $g \circ f : A \to C$ is also a surjection.

Notice that the conditions for a composite function are satisfied, so the composite function $g \circ f : A \to C$ exists. Now suppose that $c$ is any element in $C$. Then, since $g$ is surjective, there exists an element $b \in B$ such that $g(b) = c$; and as $f$ is surjective, there exists an element $a \in A$ such that $f(a) = b$. Thus $(g \circ f)(a) = g(f(a)) = g(b) = c$. So $g \circ f : A \to C$ is a surjection.

### ■ *Example 10.3.2*

Suppose that $f : A \to B$ is an injection and $g : B \to C$ is a surjection. Can you deduce that (1) the function $g \circ f : A \to C$ is a surjection, or (2) that the function $g \circ f : A \to C$ is an injection?

(1) You cannot deduce that $g \circ f : A \to C$ is a surjection. Suppose that $f : \{a,b\} \to \{p,q,r\}$ is given by $f(a) = p$ and $f(b) = q$. Then $f$ is an injection. Suppose that $g : \{p,q,r\} \to \{s,t\}$ where $g(p) = g(q) = s$ and $g(r) = t$. Then $g$ is a surjection. Now, $(g \circ f) : \{a,b\} \to \{s,t\}$ is given by $(g \circ f)(a) = s$ and $(g \circ f)(b) = s$. So there is no $x \in A$ such that $(g \circ f)(x) = t$, so $g \circ f : A \to C$ is not a surjection.

(2). Also $g \circ f : A \to C$ is not an injection, because $a$ and $b$ are distinct elements of $A$ mapping to the element $s$ in $C$ under $f$.

This section ends with a proof of an obvious-looking theorem.

**Theorem 29**: For any function $f : A \to A$, $I_A \circ f = f = f \circ I_A$.

**Proof**: For $a \in A$, $(I_A \circ f)(a) = I_A(f(a)) = f(a)$.

Also $(f \circ I_A)(a) = f(I_A(a)) = f(a)$. ∎

## 10.4 INVERSE FUNCTIONS

Here is an important question about functions. Suppose that $f : A \to B$ is a function. Under what circumstances is there a reverse function $g$ from $B$ to $A$ which has the opposite effect to $f$? That is, under what conditions does there exist a function $g : B \to A$ such that $g(b) = a$ whenever $f(a) = b$?

The answer is that there will be a function $g : B \to A$ which reverses the effect of $f : A \to B$, if, and only if, $f : A \to B$ is a bijection.

*Proving this apparently straightforward statement is long and detailed. If it would make you feel more comfortable, jump to the end of the section, and come back to fill in the details later.*

**Theorem 30** $f : A \to B$ is a bijection if, and only if, there exists a function $g : B \to A$ with $g(f(a)) = a$, for all $a \in A$, and $f(g(b)) = b$, for all $b \in B$.

**Proof**: Consider the following two statements.

(1)  There exists a function $g : B \to A$ such that $g(f(a)) = a$, for all $a \in A$ and $f(g(b)) = b$, for all $b \in B$.

(2)  $f : A \to B$ is a bijection.

*The proof will come in two stages. The first stage is to show that if statement (1) is true, then statement (2) is true. Thus you need to show that f is injective and surjective.*

Suppose that statement (1) is true.

*Injection*. If $f(x) = f(y)$, then $g(f(x)) = g(f(y))$, so, by the first part of the hypothesis of statement 1, $x = y$. Hence $f : A \to B$ is injective.

*Surjection.* If $b \in B$, then, from the second part of the hypothesis of statement (1), $b = f(g(b))$ is the image under $f$ of $g(b) \in A$. Hence $f : A \to B$ is surjective.

Hence $f : A \to B$ is bijective.

> *The second stage consists of showing that if statement (2) is true, then statement (1) is true. The first part of this stage consists in showing that the function g is well-defined.*

Suppose that $f : A \to B$ is a bijection.

*Well-defined.* Then, for every $b \in B$, there is at least one $a \in A$ such that $f(a) = b$ because $f$ is surjective. But, given $b$, there cannot be more than one $a \in A$ such that $f(a) = b$ because $f$ is injective. Therefore, specifying $g(b)$ to be that unique $a \in A$ such that $f(a) = b$ gives a well-defined function $g : B \to A$.

> *So g is well defined. Now you have to show that the function g has the required properties.*

Moreover, for $b \in B$, $f(g(b)) = f(a) = b$.

Also, for $a \in A$, if you write $b = f(a)$, then $g(b)$ is, by construction, equal to $a$, so $g(f(a)) = a$. This concludes the proof of the theorem. ∎

Theorem 30 tells you that $f : A \to B$ is a bijection if, and only if, there is a function $g : B \to A$ such that $g \circ f = I_A$ and $f \circ g = I_B$. But what can you conclude from all this? You can now define an inverse function.

**Definition**: The function $g : B \to A$ of Theorem 30 is called an **inverse** for $f$.

There are still two more results to prove in this section.

**Theorem 31**: Let $f : A \to B$ be a bijection. Then the function $g : B \to A$ with $g(f(a)) = a$, for all $a \in A$, and $f(g(b)) = b$, for all $b \in B$ is (1) a bijection, and (2) uniquely determined by $f$.

**Proof**: The function $f$ plays the role of an inverse for $g$, so Theorem 30 can be applied to $g$. This shows that $g$ is a bijection.

> Remember in proving part (2), that two functions are equal if they
> have the same domain, co-domain, and have the same effect on
> every element of the domain.

To prove the uniqueness, suppose that $g : B \to A$ and $h : B \to A$ are
both inverses of $f$. Then $f(g(b)) = b = f(h(b))$ for all $b \in B$. Therefore,
since $f$ is injective, $g(b) = h(b)$ for all $b \in B$. Therefore $g = h$. ∎

You can now talk about *the* inverse $f^{-1}$ of a bijection $f : A \to B$. This
is the function $f^{-1} : B \to A$ such that $f \circ f^{-1} = I_B$ and $f^{-1} \circ f = I_A$.

## 10.5 ASSOCIATIVITY OF FUNCTIONS

Suppose that $f : A \to B$, $g : B \to C$ and $h : C \to D$ are three functions.
Then you can combine them to form composite functions in two ways:
you could combine $f$ and $g$ first to get $g \circ f$ and then combine this with
$h$ to get the function $h \circ (g \circ f) : A \to D$; or you could work the other
way round to get $(h \circ g) \circ f : A \to D$. You probably suspect that it
doesn't matter which way you write it – the result will be the same.
You are right!

To prove it, you need to go back to the definitions in Section 10.2.

**Theorem 32**: Let $f : A \to B$, $g : B \to C$ and $h : C \to D$ be functions.
Then $(h \circ g) \circ f = h \circ (g \circ f)$.

**Proof:** Using the definition of composite function on $\left[(h \circ g) \circ f\right](x)$
gives $\left[(h \circ g) \circ f\right](x) = (h \circ g)(f(x)) = h\left[g(f(x))\right]$. In a similar way
$\left[h \circ (g \circ f)\right](x)$ gives $\left[h \circ (g \circ f)\right](x) = h\left[(g \circ f)(x)\right] = h\left[g(f(x))\right]$. As
the domains and co-domains of $(h \circ g) \circ f$ and $h \circ (g \circ f)$ are the same
and $\left[(h \circ g) \circ f\right](x) = \left[h \circ (g \circ f)\right](x)$ for any $x$ in $A$, it follows that the
functions $(h \circ g) \circ f$ and $h \circ (g \circ f)$ are equal. ∎

It is also easy to see, but tedious to write out, a proof that a kind of
generalised associative law holds for composite functions. Thus, if $p$, $q$,
$r$ and $s$ are four functions with compatible domains and co-domains so
that they can be composed together, you can talk about the function
$s \circ r \circ q \circ p$ without brackets.

## 10.6 INVERSE OF A COMPOSITE FUNCTION

It is useful now to introduce a collapsed form of the usual diagram for sets and functions. Let $f : A \to B$ and $g : B \to C$ be functions. These functions are illustrated in Fig. 10.2, together with $g \circ f : A \to C$.

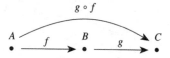

**Fig. 10.2** *The function* $g \circ f : A \to C$

Now suppose that the functions $f : A \to B$ and $g : B \to C$ are both bijections, so that the inverse functions $f^{-1} : B \to A$ and $g^{-1} : C \to B$ both exist. What can you say about the inverse of the composite function $g \circ f : A \to C$? Fig. 10.3 helps to make the situation clear.

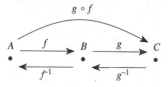

**Fig. 10.3** *The inverse of* $g \circ f : A \to C$

From Fig. 10.3 you can surmise that the inverse of the function $g \circ f$ is $f^{-1} \circ g^{-1}$. Here is a proof.

> *To prove this result you need to go back to Section 10.4 and check the details. You need to check that* $\left(f^{-1} \circ g^{-1}\right) \circ (g \circ f) = I_A$ *and* $(g \circ f) \circ \left(f^{-1} \circ g^{-1}\right) = I_C$

**Theorem 33:** Let the two functions $f : A \to B$ and $g : B \to C$ be bijections. Then the inverse of the function $g \circ f : A \to C$ exists and is the function $\left(f^{-1} \circ g^{-1}\right) : C \to A$. Furthermore $g \circ f$ and $f^{-1} \circ g^{-1}$ are both bijections.

**Proof:** First you need to show that $\left(f^{-1} \circ g^{-1}\right) \circ (g \circ f) = I_A$. Then you need to show that $(g \circ f) \circ \left(f^{-1} \circ g^{-1}\right) = I_C$.

First,

$$\left(f^{-1} \circ g^{-1}\right) \circ (g \circ f) = f^{-1} \circ \left(g^{-1} \circ (g \circ f)\right)$$

$$= f^{-1} \circ \left(\left(g^{-1} \circ g\right) \circ f\right)$$

$$= f^{-1} \circ \left(I_B \circ f\right)$$

$$= f^{-1} \circ f \qquad \text{using Theorem 29}$$

$$= I_A$$

Secondly,

$$(g \circ f) \circ \left(f^{-1} \circ g^{-1}\right) = g \circ \left(f \circ \left(f^{-1} \circ g^{-1}\right)\right)$$

$$= g \circ \left(\left(f \circ f^{-1}\right) \circ g^{-1}\right)$$

$$= g \circ \left(I_B \circ g^{-1}\right)$$

$$= g \circ g^{-1} \qquad \text{again using Theorem 29}$$

$$= I_C$$

Therefore $f^{-1} \circ g^{-1}$ is the inverse of $g \circ f$.

Also it follows from Theorem 30 that $g \circ f$ and $f^{-1} \circ g^{-1}$ are bijections. ∎

*You can think of Theorem 33 in terms of doing and un-doing certain actions. For example, if you want to reverse the effect of putting on your socks and then putting on your climbing boots, you would first take off your boots and then take off your socks.*

## 10.7 THE BIJECTIONS FROM A SET TO ITSELF

You have probably guessed by now that the set of bijections from a set $A$ to itself forms a group. Here is a proof.

**Theorem 34**: The set of bijections from a set $A$ to itself together with the operation of composition is a group.

**Proof:** Suppose that $f : A \rightarrow A$ and $g : A \rightarrow A$ are bijections. The proof that $f \circ g$ is a bijection follows immediately from Theorem 33.

Let $f$, $g$ and $h$ be bijections $A \to A$. Then, by Theorem 32, $f \circ (g \circ h) = (f \circ g) \circ h$. So bijections are associative under composition of functions.

The identity function $I_A : A \to A$ is clearly a bijection, and, from Theorem 29, $I_A \circ f = f \circ I_A = f$ for any bijection $f : A \to A$. This identity bijection, $I_A$, therefore fulfils the role of identity in the group.

Finally, from Theorems 30 and 31, as $f : A \to A$ is a bijection, the function $f^{-1} : A \to A$ exists, is a bijection and $f^{-1} \circ f = f \circ f^{-1} = I_A$.

Hence all the group axioms are satisfied, so the set of bijections $A \to A$ together with the operation of composition is a group. ∎

*You should be aware that the word 'identity' is used in two distinct ways in the middle of this proof. The identity function from A to itself fulfils the role of the identity element in the group.*

## WHAT YOU SHOULD KNOW

- What the composition of two functions means, the notation for it, and the conditions under which it exists.

- What an inverse function is, conditions for the inverse function to exist, and how to find the inverse of a product of functions.

- Composition of functions is associative.

- The set of bijections from a set $A$ to itself together with the operation of composition is a group.

## EXERCISE 10

**1** Prove that, if $f : A \to B$ and $g : B \to A$ are such that $g \circ f = I_A$, then $f$ is a injection and $g$ is a surjection.

**2** In question 2 of Exercise 9 you decided whether each of the following definitions was the definition of a function and, if so, whether it was injective and whether it was surjective. Use your answer to decide which of the functions have inverses, and give these inverses.

(1)  $f(x) = x^2$
(2)  $f(x) = x^3$
(3)  $f(x) = 1/x$

(4)   $f(x) = \cos x$

(5)   $f(x) = \tan x$

(6)   $f(x) = e^x$

(7)   $f(x) = |x|$

(8)   $f(x) = \text{int } x$  ( int $x$ is the largest integer $\leq x$.)

(9)   $f(x) = \sqrt{x}$

(10)   $f(x) = x + 1$

(11)   $f(x) = \sin^{-1} x$

(12)   $f(x)$ is the smallest real number greater than $x$.

**3**   Show that the function $g : \mathbf{Z}^+ \to \mathbf{Z}$ defined by

$$g(n) = \begin{cases} n/2, \text{ if } n \text{ is even} \\ (1-n)/2, \text{ if } n \text{ is odd} \end{cases}$$

is inverse to the function $f : \mathbf{Z} \to \mathbf{Z}^+$ of question 8 of Exercise 9.

**4**   Mark each of the following statements true or false.

(1)   The inverse of $f \circ g$ is $f^{-1} \circ g^{-1}$.

(2)   If $f$ is a bijection, then $f$ is an injection.

(3)   An injection is a bijection if, and only if, it is a surjection.

(4)   An injection from a finite set to itself is also a surjection.

(5)   An injection from an infinite set to itself is also a surjection.

**5**   Let $A$ be a set, and $X$ be any subset of $A$. From Theorem 34 the set $B$ of bijections $A \to A$ forms a group under composition of functions. Prove that the subset $H = \{ f \in B : f(X) = X \}$ is a subgroup of $B$.

Notice that the fact that $f(X) = X$ does not mean that $f(x) = x$ for all $x \in X$. If $f(X) = X$, then for all $x \in X$, $f(x)$ is an element of $X$, but $f(x)$ need not be equal to the particular element $x$.

# 11

# *Isomorphisms*

## 11.1 INTRODUCTION

In working Chapters 6, 7 and Chapter 8 you must have noticed the similarity between some of the group tables which you constructed.

For example, the group consisting of $\{1, -1\}$ under multiplication is very similar to the subgroup of $D_3$ consisting of $\{I, X\}$. Figures 11.1 and 11.2 show these two groups; you can see that they have the same structure.

|     | 1   | −1  |
| --- | --- | --- |
| 1   | 1   | −1  |
| −1  | −1  | 1   |

|     | $I$ | $X$ |
| --- | --- | --- |
| $I$ | $I$ | $X$ |
| $X$ | $X$ | $I$ |

**Fig. 11.1**                    **Fig. 11.2**

If you replace 1 by $I$ and −1 by $X$, the tables become identical; the only difference is the names of the elements. This is an example of an isomorphism. Here are two more examples.

## ■ *Example 11.1.1*

The group $(\mathbf{Z}_4, +)$ is illustrated in Fig. 11.3 and the group $(\{1, i, -1, -i\}, \times)$ is shown in Fig. 11.4.

|   | 0 | 1 | 2 | 3 |
|---|---|---|---|---|
| 0 | 0 | 1 | 2 | 3 |
| 1 | 1 | 2 | 3 | 0 |
| 2 | 2 | 3 | 0 | 1 |
| 3 | 3 | 0 | 1 | 2 |

|    | 1  | $i$ | -1 | $-i$ |
|----|----|----|----|----|
| 1  | 1  | $i$ | -1 | $-i$ |
| $i$  | $i$  | -1 | $-i$ | 1  |
| -1 | -1 | $-i$ | 1  | $i$  |
| $-i$ | $-i$ | 1  | $i$  | -1 |

**Fig. 11.3**                    **Fig. 11.4**

Once again you can see that these tables are identical in structure. The only difference is in the names of the elements. If you replace 0 by 1, 1 by $i$, 2 by –1 and 3 by $-i$, the tables become identical.

> It is not always so easy to see when two group tables are identical. For example, the tables may be too big to write out, or the information may be concealed in some way.

## ■ *Example 11.1.2*

|   | 2 | 4 | 6 | 8 |
|---|---|---|---|---|
| 2 | 4 | 8 | 2 | 6 |
| 4 | 8 | 6 | 4 | 2 |
| 6 | 2 | 4 | 6 | 8 |
| 8 | 6 | 2 | 8 | 4 |

|   | 0 | 1 | 2 | 3 |
|---|---|---|---|---|
| 0 | 0 | 1 | 2 | 3 |
| 1 | 1 | 2 | 3 | 0 |
| 2 | 2 | 3 | 0 | 1 |
| 3 | 3 | 0 | 1 | 2 |

**Fig. 11.5**

Consider now the group in Example 5.2.1, and $(\mathbf{Z}_4, +)$. Are these two groups, shown in Fig. 11.5, structurally the same?

The situation is not quite so straightforward. At first sight you might say that the two groups do not have the same structure, but wait! If you first interchange the rows 2 ‖ 4 8 2 6 and 6 ‖ 2 4 6 8 and then interchange the columns headed by 2 and 6 in the left-hand table of Fig. 11.5 you would get Fig. 11.6; if you now interchange the rows 2 ‖ 2 8 4 6 and 4 ‖ 4 6 8 2 and then interchange the columns headed by 2 and 4 you get Fig. 11.7.

| | 6 | 4 | 2 | 8 |
|---|---|---|---|---|
| 6 | 6 | 4 | 2 | 8 |
| 4 | 4 | 6 | 8 | 2 |
| 2 | 2 | 8 | 4 | 6 |
| 8 | 8 | 2 | 6 | 4 |

| | 6 | 2 | 4 | 8 |
|---|---|---|---|---|
| 6 | 6 | 2 | 4 | 8 |
| 2 | 2 | 4 | 8 | 6 |
| 4 | 4 | 8 | 6 | 2 |
| 8 | 8 | 6 | 2 | 4 |

**Fig. 11.6**                    **Fig. 11.7**

You can see now that Figs. 11.7 and 11.3 do have the same structure, with $6 \to 0$, $2 \to 1$, $4 \to 2$ and $8 \to 3$. So, to decide whether or not two tables have the same structure you may have to re-arrange them.

Here is an example where the two groups are not isomorphic.

## ■ *Example 11.1.3*

The first group is $Z_2 \times Z_2$, and the second is $(Z_4, +)$. If you look at Fig. 11.8 overleaf, at the terms on the diagonal, you will see that in $Z_2 \times Z_2$ each term multiplied by itself gives $(0,0)$, the identity element in $Z_2 \times Z_2$, whereas in $Z_4$ the same is not true. This is an example of a different structure, because no amount of re-arranging the order of the elements or renaming the elements in Fig. 11.9 can make all the elements on the diagonal equal to 0, the identity element in $Z_4$. This is because the elements in the diagonal are always the 'squares' of the elements in the group, and in $Z_4$, $1 + 1 = 2$ and $3 + 3 = 2$.

|   | (0,0) | (1,0) | (0,1) | (1,1) |
|---|---|---|---|---|
| (0,0) | (0,0) | (1,0) | (0,1) | (1,1) |
| (1,0) | (1,0) | (0,0) | (1,1) | (0,1) |
| (0,1) | (0,1) | (1,1) | (0,0) | (1,0) |
| (1,1) | (1,1) | (0,1) | (1,0) | (0,0) |

|   | 0 | 1 | 2 | 3 |
|---|---|---|---|---|
| 0 | 0 | 1 | 2 | 3 |
| 1 | 1 | 2 | 3 | 0 |
| 2 | 2 | 3 | 0 | 1 |
| 3 | 3 | 0 | 1 | 2 |

**Fig. 11.8** $\mathbf{Z}_2 \times \mathbf{Z}_2$       **Fig. 11.9** $(\mathbf{Z}_4, +)$

So $\mathbf{Z}_2 \times \mathbf{Z}_2$ does not have the same structure as $\mathbf{Z}_4$.

## 11.2 ISOMORPHISM

The literal meaning of this word is 'equal structure', and it is used in this sense mathematically.

Thinking informally, if two groups $G$ and $H$ are to have the same structure, you would expect them to be the same size. So it is a good idea to insist that a bijection exists between the elements of $G$ and $H$.

> You have seen in Example 11.1.3 that simply to have a bijection is not sufficient. It is almost as if you go into a car showroom and you have to decide between two apparently identical cars. It isn't sufficient that they have the same parts; the parts have to fit together and work together in the same way. If there was simply a bijection between the parts, then one car might be assembled, ready to drive, and the other could be a heap of parts on the floor.

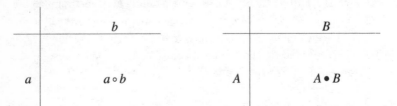

|   | $b$ |
|---|---|
| $a$ | $a \circ b$ |

|   | $B$ |
|---|---|
| $A$ | $A \bullet B$ |

**Fig. 11.10** *This leads to a definition of isomorphism.*

However, you can guess at the important extra condition by looking carefully at the partially completed Fig. 11.10. Suppose that you have defined a bijection $f$ between $G$ and $H$, so that $f(a) = A$ and $f(b) = B$. Then, if the groups have the same structure, the product $a \circ b$ must correspond to the product $A \bullet B$; that is $f(a \circ b) = A \bullet B$. And, moreover, this must be true for all possible elements $a$ and $b$.

**Definition:** Two groups $(G, \circ)$ and $(H, \bullet)$ are **isomorphic** if there is a bijection $f : G \to H$ such that for each $x \in G$ and $y \in G$, $f(x \circ y) = f(x) \bullet f(y)$. The function $f$ is called an **isomorphism**. The notation $G \cong H$ will be used to denote that $G$ is isomorphic to $H$.

Often the notation used is simply the ordinary multiplication notation in both groups. The condition $f(x \circ y) = f(x) \bullet f(y)$ then becomes $f(xy) = f(x)f(y)$.

There are some immediate results which follow from this definition. For example, since you are really re-naming the elements, you would be very surprised indeed if the image of the identity element in $G$ were not the identity element in $H$. Similarly, if $x \in G$, you would expect $x^{-1}$ to map into the element which is the inverse of the image $f(x)$ in $H$.

**Theorem 35:** If $f : G \to H$ is an isomorphism of $G$ and $H$, and $e$ is the identity in $G$, then $f(e)$ is the identity in $H$. Similarly, if $x \in G$, then $f\left(x^{-1}\right)$ is the inverse of $f(x)$ in $H$.

**Proof:**

*On the way expect to use the relation $f(xy) = f(x)f(y)$.*

Let $e$ be the identity in $G$. Then $f(e)f(e) = f(ee) = f(e)$, so therefore $\left(f(e)\right)^{-1} f(e) f(e) = \left(f(e)\right)^{-1} f(e)$ which shows that $e_H f(e) = e_H$ or $f(e) = e_H$.

*To prove that $f\left(x^{-1}\right)$ is the inverse of $f(x)$, you must prove that $f\left(x^{-1}\right)f(x) = f(x)f\left(x^{-1}\right) = e_H$. So start with $f\left(x^{-1}\right)f(x)$.*

Using $f(xy) = f(x)f(y)$, $f\left(x^{-1}\right)f(x) = f\left(x^{-1}x\right) = f(e) = e_H$. Also, $f(x)f\left(x^{-1}\right) = f\left(xx^{-1}\right) = f(e) = e_H$. This proves the result. ∎

## 11.3 PROVING TWO GROUPS ARE ISOMORPHIC

To prove that two groups are isomorphic, you need to produce a bijection, and show that the relation $f(xy) = f(x)f(y)$ holds.

If the groups are small, you can do this efficiently by writing out the groups' tables in a corresponding way. This is, in effect, producing the bijection, and verifying, case by case, that $f(xy) = f(x)f(y)$.

However, usually it will be more practical to specify the isomorphism differently. Here is an example.

### ■ *Example 11.3.1*

Prove that $\mathbf{Z}_6 \cong \mathbf{Z}_2 \times \mathbf{Z}_3$.

Define a function $f : \mathbf{Z}_6 \to \mathbf{Z}_2 \times \mathbf{Z}_3$ by $f([a]_6) = ([a]_2, [a]_3)$.

You need to show that this function is well defined. The image $([a]_2, [a]_3)$ is specified in terms of $a$, which is just one of many possible 'representatives' for the class $[a]_6$. To show $f$ is well defined you need to show that $([a]_2, [a]_3)$ is independent of the particular choice of $a$.

*Well defined*. Suppose that $a, b \in \mathbf{Z}$ such that $[a]_6 = [b]_6$. You need to show $([a]_2, [a]_3) = ([b]_2, [b]_3)$. This is true because if $[a]_6 = [b]_6$, then $a \equiv b \pmod 6$, so 6 divides $(a - b)$. Therefore 2 divides $(a - b)$ and 3 divides $(a - b)$, and therefore, $a \equiv b \pmod 2$ and $a \equiv b \pmod 3$. It follows that $[a]_2 = [b]_2$ and $[a]_3 = [b]_3$, so $([a]_2, [a]_3) = ([b]_2, [b]_3)$.

*Injection*. If $f([a]_6) = f([b]_6)$, then $([a]_2, [a]_3) = ([b]_2, [b]_3)$, so $[a]_2 = [b]_2$ and $[a]_3 = [b]_3$. Therefore $a \equiv b \pmod 2$ and $a \equiv b \pmod 3$. Thus 2 divides $a - b$ and 3 divides $a - b$, so 6 divides $a - b$. Therefore $a \equiv b \pmod 6$ and $[a]_6 = [b]_6$.

*Surjection*. $\mathbf{Z}_6$ and $\mathbf{Z}_2 \times \mathbf{Z}_3$ both have six elements, and $f$ is an injection. Therefore, from Theorem 27, $f$ is a surjection.

Finally,

$$f\big([a]_6+[b]_6\big) = f\big([a+b]_6\big)$$
$$= \big([a+b]_2,[a+b]_3\big)$$
$$= \big([a]_2+[b]_2,[a]_3+[b]_3\big)$$
$$= \big([a]_2,[a]_3\big) + \big([b]_2,[b]_3\big)$$
$$= f\big([a]_6\big) + f\big([b]_6\big)$$

proving that the two groups are isomorphic. ∎

The next theorem shows that, up to isomorphism, there is just one cyclic group of any given order.

**Theorem 36**: (1) Every infinite cyclic group is isomorphic to $(\mathbf{Z},+)$.

(2) Every cyclic group of finite order $n$ is isomorphic to $(\mathbf{Z}_n,+)$.

**Proof**: (1) Let $A$ be an infinite cyclic group, and let $a$ be a generator for $A$. Define $f : \mathbf{Z} \to A$ by $f(r)=a^r$.

*Surjection.* Every element of $A$ is a power of $a$.

*Injection.* If $f(r)=f(s)$, then $a^r=a^s$, which, by Theorem 18, part (1), gives $r=s$. Therefore, $f$ is an injection.

Finally $f(r+s)=a^{r+s}=a^ra^s=f(r)f(s)$. So $f$ is an isomorphism.

(2) Let $A$ be a finite cyclic group of order $n$ and let $a$ be a generator for $A$. Define $f : \mathbf{Z}_n \to A$, by $f\big([r]\big)=a^r$.

*Well defined.* To show that this function is well defined, you must show that if $[r]=[s]$ then $f\big([r]\big)=f\big([s]\big)$, that is, $a^r=a^s$. If $[r]=[s]$, then $r \equiv s \,(\mathrm{mod}\,n)$, so $a^r=a^s$, by Theorem 18, part (2).

*Surjection.* As every element in $A$ is a power of $a$, $f$ is a surjection.

*Injection.* By Theorem 27, since $\big|\mathbf{Z}_n\big|=\big|A\big|$ and $f$ is a surjection, $f$ is an injection.

Finally, $f\big([r]+[s]\big)=f\big([r+s]\big)=a^{r+s}=a^ra^s=f\big([r]\big)f\big([s]\big)$. So $f$ is an isomorphism. ∎

Theorem 36 shows that all cyclic groups of a given order are isomorphic to each other. This is often described by saying that there is only one cyclic group of a given order, up to isomorphism. In this sense you can talk about 'the' infinite cyclic group and 'the' finite cyclic group of order $n$.

In this book, the notation $\mathbf{Z}$ will be used for the general infinite cyclic group as well as for the specific group of integers under addition. Similarly, the notation $\mathbf{Z}_n$ will be used for the general cyclic group of order $n$ as well as for the specific group of residue classes of integers modulo $n$ under addition. Occasionally, and especially in Chapter 15, where there are explicit calculations carried out in groups written in multiplicative notation, the cyclic group of order $n$ will be denoted by $C_n = \left\{ e, a, a^2, \ldots, a^{n-1} \right\}$.

## 11.4 PROVING TWO GROUPS ARE NOT ISOMORPHIC

If two groups $G$ and $H$ do not have the same order, that is, the same number of elements, then there cannot be a bijection between them, so the groups will not be isomorphic.

If there is a bijection between the groups $G$ and $H$, then you have to look for something different about their algebraic structures. One good place to look is at the number of solutions of $x^n = e$, where $n$ is fixed, in each of $G$ and $H$. If these two numbers are different, then $G$ and $H$ are not isomorphic.

To see why this is so, let $f : G \to H$ be an isomorphism. Writing $T(G) = \left\{ x \in G : x^n = e_G \right\}$ and $T(H) = \left\{ x \in H : x^n = e_H \right\}$, suppose that $x \in T(G)$. Then $\left( f(x) \right)^n = \overbrace{f(x) \ldots f(x)}^{n \text{ times}} = f\left( x^n \right) = f(e_G) = e_H$, so that $f(x) \in T(H)$. Therefore $f\left( T(G) \right) \subseteq T(H)$.

Suppose now that $h \in T(H)$. Since $f$ is a bijection, $h$ is the image of some element $g \in G$. Then $e_H = h^n = \overbrace{f(g) \ldots f(g)}^{n \text{ times}} = f\left( g^n \right)$, showing that $e_H$ is the image of $g^n \in G$. But, since $f$ is a bijection, $e_H$ is the image of just one element in $G$, and, by Theorem 35, $e_H$ is the image of $e_G$. Therefore $g^n = e_G$ so $g \in T(G)$. Therefore $T(H) \subseteq f(T(G))$.

Therefore $f(T(G)) \subseteq T(H)$ and $T(H) \subseteq f(T(G))$, so $f(T(G)) = T(H)$.

It follows from this that the group $(\mathbf{C}^*, \times)$ is not isomorphic to the group $(\mathbf{R}^*, \times)$ under multiplication. The identity is 1: there are four solutions to the equation $x^4 = 1$ in $\mathbf{C} - \{0\}$, but only two solutions in $(\mathbf{R}^*, \times)$.

## 11.5 FINITE ABELIAN GROUPS

It is now possible to prove some results about finite abelian groups.

**Definition**: Let $A$ be an abelian group. A subset $S = \{a_1, a_2, \ldots, a_n\}$ of $A$ is said to be a **generating set** for $A$ if every element of $A$ is a product of elements of $S$.

Since $A$ is abelian, it follows immediately from this definition that every element of $A$ will be of the form $a_1^{r_1} a_2^{r_2} \ldots a_n^{r_n}$ where $r_1, r_2, \ldots, r_n \in \mathbf{Z}$.

This follows because, as $A$ is abelian you can regroup any product such as $a_1 a_2 a_3 a_1 a_2 a_2 a_3 a_1$ in the form $a_1^3 a_2^3 a_3^2$.

You can write $A = \langle a_1, a_2, \ldots, a_n \rangle$, or $A = \langle S \rangle$, to show that $A$ is generated by $S = \{a_1, a_2, \ldots, a_n\}$. This is an obvious extension of the $\langle \ \rangle$ notation which first arose in Section 6.4, after Theorem 22. A generating set $S$ is a **minimal generating set** if there is no proper subset of $S$ which is also a generating set for $A$.

**Theorem 37**: Let $A$ be a finite group in which every element (other than the identity) has order 2, and let $\{a_1, a_2, \ldots, a_n\}$ be a minimal generating set for $A$. Then (1) $A$ is abelian and (2) every element of $A$ can be written uniquely in the form $a_1^{\varepsilon_1} a_2^{\varepsilon_2} \ldots a_n^{\varepsilon_n}$, where, for each $i = 1, 2, \ldots, n$, $\varepsilon_i = 0$ or 1.

**Proof**: (1) Let $a, b \in A$. As every element of $A$ other than the identity $e$ has order 2, every element $x$ of $G$ satisfies $x^2 = e$. So $abab = (ab)^2 = e$ and it follows from this that $ba = a^{-1}b^{-1}$. But also

$a^2 = e$ and $b^2 = e$. Therefore $a^{-1} = a$ and $b^{-1} = b$, and so $ab = a^{-1}b^{-1} = ba$. Thus $A$ is abelian.

(2) *Existence.* By Theorem 18, part (2), for each $i$, every power of $a_i$ is equal to $e$ or $a_i$. Therefore, as $\{a_1, a_2, \ldots, a_n\}$ is a generating set for $A$, every element of $A$ is of the form $a_1^{\varepsilon_1} a_2^{\varepsilon_2} \ldots a_n^{\varepsilon_n}$ where, for each $i = 1, 2, \ldots, n$, $\varepsilon_i = 0$ or 1.

*Uniqueness.* Let $x$ be an element of $A$ and suppose that $x$ can be expressed in two ways, $a_1^{\varepsilon_1} a_2^{\varepsilon_2} \ldots a_n^{\varepsilon_n} = x = a_1^{\delta_1} a_2^{\delta_2} \ldots a_n^{\delta_n}$, where for each $i = 1, 2, \ldots, n$, $\varepsilon_i$ and $\delta_i$ are 0 or 1. Suppose that it is not the case that $\varepsilon_i = \delta_i$ for all $i = 1, 2, \ldots, n$, and let $r$ be the least value of $i$ such that $\varepsilon_i \neq \delta_i$. Then $a_1^{\varepsilon_1} a_2^{\varepsilon_2} \ldots a_{r-1}^{\varepsilon_{r-1}} = a_1^{\delta_1} a_2^{\delta_2} \ldots a_{r-1}^{\delta_{r-1}}$, and it follows that $a_r^{\varepsilon_r} \ldots a_n^{\varepsilon_n} = a_r^{\delta_r} \ldots a_n^{\delta_n}$. where $\varepsilon_r \neq \delta_r$. Therefore $a_r^{\varepsilon_r} \left( a_r^{\delta_r} \right)^{-1} = a_{r+1}^{\delta_{r+1}} \ldots a_n^{\delta_n} \left( a_{r+1}^{\varepsilon_{r+1}} \ldots a_n^{\varepsilon_n} \right)^{-1}$.

But $a_r^{\varepsilon_r} \left( a_r^{\delta_r} \right)^{-1} = a_r^{\varepsilon_r} a_r^{-\delta_r} = a_r^{\varepsilon_r - \delta_r} = a_r$, because $\varepsilon_r - \delta_r \neq 0$ and each of $\varepsilon_r$ and $\delta_r$ is equal to 0 or 1. Therefore $a_r$ can be written as a product of powers of the elements $a_i$ with $i \neq r$ and hence $S = \{a_1, \ldots, a_{r-1}, a_{r+1}, \ldots, a_n\}$ is a generating set for $A$. This contradicts the fact that $\{a_1, a_2, \ldots, a_n\}$ is a minimal generating set for $A$.

Therefore $\varepsilon_i = \delta_i$ for all $i = 1, 2, \ldots, n$. ∎

*As each of the exponents $\varepsilon_i$ is either 0 or 1, it follows that the number of elements of A is equal to $2^n$. Recall that, in question 10 of Exercise 2 you were asked to prove that a set of n elements had $2^n$ subsets. As each expression of the form $a_1^{\varepsilon_1} a_2^{\varepsilon_2} \ldots a_n^{\varepsilon_n}$ specifies a unique subset of $\{a_1, a_2, \ldots, a_n\}$; namely the subset consisting of the $a_i$ for which $\varepsilon_i = 1$, there is a bijection $f : \{B : B \subseteq \{a_1, a_2, \ldots, a_n\}\} \to A$ given by $f(B) = $ product of the elements of B.*

**Theorem 38**: Let $A$ be a finite group in which every element other than the identity has order 2. Then $A \cong \mathbf{Z}_2 \times \ldots \times \mathbf{Z}_2$, the direct product of $n$ copies of $\mathbf{Z}_2$.

**Proof**: Let $A = \{a_1, a_2, \ldots, a_n\}$ be a minimal generating set for $A$, and let $D = \mathbf{Z}_2 \times \mathbf{Z}_2 \times \ldots \times \mathbf{Z}_2$. Define the function $f : D \to A$ by $f([r_1], [r_2], \ldots, [r_n]) = a_1^{r_1} a_2^{r_2} \ldots a_n^{r_n}$.

*Well defined.* If $\left(\left[r_1\right],\left[r_2\right],\ldots,\left[r_n\right]\right)=\left(\left[s_1\right],\left[s_2\right],\ldots,\left[s_n\right]\right)$, then
$r_i \equiv s_i \pmod 2$ for all $i=1,2,\ldots,n$, so $a_i^{r_i}=a_i^{s_i}$ for all $i=1,2,\ldots,n$
by Theorem 18, part (2), and $a_1^{r_1}a_2^{r_2}\ldots a_n^{r_n}=a_1^{s_1}a_2^{s_2}\ldots a_n^{s_n}$.

*Surjection.* Every element of $A$ is of the form $a_1^{r_1}a_2^{r_2}\ldots a_n^{r_n}$, because $\{a_1,a_2,\ldots,a_n\}$ is a generating set for $A$.

*Injection.* As $D$ and $A$ have the same number of elements, namely $2^n$, and $f$ is surjective, it follows from Theorem 27 that $f$ is injective.

$$\text{Finally,} \quad f\left(\left(\left[r_1\right],\left[r_2\right],\ldots,\left[r_n\right]\right)+\left(\left[s_1\right],\left[s_2\right],\ldots,\left[s_n\right]\right)\right)$$

$$= f\left(\left(\left[r_1+s_1\right],\left[r_2+s_2\right],\ldots,\left[r_n+s_n\right]\right)\right)$$

$$= a_1^{r_1+s_1}a_2^{r_2+s_2}\ldots a_n^{r_n+s_n}$$

$$= a_1^{r_1}a_1^{s_1}a_2^{r_2}a_2^{s_2}\ldots a_n^{r_n}a_n^{s_n}$$

$$= \left(a_1^{r_1}a_2^{r_2}\ldots a_n^{r_n}\right)\left(a_1^{s_1}a_2^{s_2}\ldots a_n^{s_n}\right)$$

$$= f\left(\left[r_1\right],\left[r_2\right],\ldots,\left[r_n\right]\right)f\left(\left[s_1\right],\left[s_2\right],\ldots,\left[s_n\right]\right)$$

Therefore $f$ is an isomorphism. ∎

*Notice that the proof of this theorem has a very similar structure to the proof of Theorem 36, part (2).*

This result is a particular case of a more general theorem about finite abelian groups. We shall state this more general result without proof.

We know, from Theorem 5, that a given positive integer can be written uniquely as a product of powers of distinct prime numbers $n = p_1^{r_1}p_2^{r_2}\ldots p_k^{r_k}$ but there may be several different-looking factorisations if you allow $p_i = p_j$ for various distinct *i*s and *j*s.

For example, $360 = 2^3 \times 3^2 \times 5$. There is only one factorisation as powers of the distinct primes 2, 3 and 5, but, allowing for repetitions, there are six factorisations in all:

$$360 = 2^3 \times 3^2 \times 5$$
$$= 2^3 \times 3 \times 3 \times 5$$
$$= 2 \times 2^2 \times 3^2 \times 5$$
$$= 2 \times 2^2 \times 3 \times 3 \times 5$$
$$= 2 \times 2 \times 2 \times 3^2 \times 5$$
$$= 2 \times 2 \times 2 \times 3 \times 3 \times 5$$

Theorem 39 says that for all such different factorisations of $n$, no two of the groups $G \cong \mathbf{Z}_{p_1^{r_1}} \times \mathbf{Z}_{p_2^{r_2}} \times \ldots \times \mathbf{Z}_{p_k^{r_k}}$ are isomorphic to each other and that every abelian group of order $n$ is isomorphic to one of these groups.

**Theorem 39**: Let $G$ be a finite abelian group. Then $G \cong \mathbf{Z}_{p_1^{r_1}} \times \mathbf{Z}_{p_2^{r_2}} \times \ldots \times \mathbf{Z}_{p_n^{r_n}}$, where the $p_i$ are primes, not necessarily distinct. This direct product of cyclic groups which have orders that are powers of primes, and which is isomorphic to $G$, is unique except for a rearrangement of the factors. ∎

## ■ *Example 11.5.1*

Consider the abelian groups of order 8. The factorisations of 8 are $8 = 2 \times 2 \times 2 = 2 \times 2^2 = 2^3$. Theorem 39 says that $G \cong \mathbf{Z}_2 \times \mathbf{Z}_2 \times \mathbf{Z}_2$, $G \cong \mathbf{Z}_2 \times \mathbf{Z}_4$ or $G \cong \mathbf{Z}_8$, and that these groups are different, that is, no two of them are isomorphic.

## ■ *Example 11.5.2*

Consider the abelian groups of order 12. The factorisations of 12 are $12 = 2 \times 2 \times 3 = 2^2 \times 3$. Theorem 39 says that $G \cong \mathbf{Z}_2 \times \mathbf{Z}_2 \times \mathbf{Z}_3$ or $G \cong \mathbf{Z}_4 \times \mathbf{Z}_3$, and that these two groups are not isomorphic.

## ■ *Example 11.5.3*

Here are the abelian groups of order 360.

$$\mathbf{Z}_8 \times \mathbf{Z}_9 \times \mathbf{Z}_5$$
$$\mathbf{Z}_8 \times \mathbf{Z}_3 \times \mathbf{Z}_3 \times \mathbf{Z}_5$$
$$\mathbf{Z}_2 \times \mathbf{Z}_4 \times \mathbf{Z}_9 \times \mathbf{Z}_5$$
$$\mathbf{Z}_2 \times \mathbf{Z}_4 \times \mathbf{Z}_3 \times \mathbf{Z}_3 \times \mathbf{Z}_5$$
$$\mathbf{Z}_2 \times \mathbf{Z}_2 \times \mathbf{Z}_2 \times \mathbf{Z}_9 \times \mathbf{Z}_5$$
$$\mathbf{Z}_2 \times \mathbf{Z}_2 \times \mathbf{Z}_2 \times \mathbf{Z}_3 \times \mathbf{Z}_3 \times \mathbf{Z}_5$$

*To see the connection between Theorem 38 and Theorem 39, look at the orders of elements. If $G \cong H \times \mathbf{Z}_d$, then $G$ has an element of order $d$, namely $(e_H, [1]_d)$. So if all the elements of $G$ other than the identity have order 2, then none of the $\mathbf{Z}_{p^r}$ factors can have $p^r$ greater than 2. So they are all $\mathbf{Z}_2$.*

Now, from a generalisation of Example 11.3.1, namely that if $m$ and $n$ are relatively prime, then $\mathbf{Z}_m \times \mathbf{Z}_n \cong \mathbf{Z}_{mn}$, by grouping various factors, you can prove Theorem 40.

*In Exercise 11, question 10, you are asked to prove that $\mathbf{Z}_m \times \mathbf{Z}_n \cong \mathbf{Z}_{mn}$ and a proof of this result appears in the answers.*

**Theorem 40**: Let $G$ be a finite abelian group. Then $G \cong \mathbf{Z}_{d_1} \times \mathbf{Z}_{d_2} \times \ldots \times \mathbf{Z}_{d_n}$, where $d_i$ divides $d_{i+1}$ for each $i = 1, 2, \ldots, n-1$. The numbers $d_1, d_2, \ldots, d_n$ are unique. ∎

## ■ *Example 11.5.4*

Theorem 40 says that the two groups of order 12 identified in Example 11.5.2 can be written as $G \cong \mathbf{Z}_2 \times \mathbf{Z}_2 \times \mathbf{Z}_3 \cong \mathbf{Z}_2 \times \mathbf{Z}_6$ and $G \cong \mathbf{Z}_4 \times \mathbf{Z}_3 \cong \mathbf{Z}_{12}$.

## ■ *Example 11.5.5*

Return to the abelian groups of order 360 in Example 11.5.3.

*You can see from the patterns of the factors on the right of the double lines (overleaf) how the direct products on the left arise.*

$$\mathbf{Z}_8 \times \mathbf{Z}_9 \times \mathbf{Z}_5 \cong \mathbf{Z}_{360} \quad \left\| \begin{array}{l} 2^3 \\ 3^2 \\ 5 \end{array} \right.$$

$$\mathbf{Z}_8 \times \mathbf{Z}_3 \times \mathbf{Z}_3 \times \mathbf{Z}_5 \cong \mathbf{Z}_3 \times \mathbf{Z}_{120} \quad \left\| \begin{array}{l} 2^3 \\ 3 \times 3 \\ 5 \end{array} \right.$$

$$\mathbf{Z}_2 \times \mathbf{Z}_4 \times \mathbf{Z}_9 \times \mathbf{Z}_5 \cong \mathbf{Z}_2 \times \mathbf{Z}_{180} \quad \left\| \begin{array}{l} 2^2 \times 2 \\ 3^2 \\ 5 \end{array} \right.$$

$$\mathbf{Z}_2 \times \mathbf{Z}_4 \times \mathbf{Z}_3 \times \mathbf{Z}_3 \times \mathbf{Z}_5 \cong \mathbf{Z}_6 \times \mathbf{Z}_{60} \quad \left\| \begin{array}{l} 2^2 \times 2 \\ 3 \times 3 \\ 5 \end{array} \right.$$

$$\mathbf{Z}_2 \times \mathbf{Z}_2 \times \mathbf{Z}_2 \times \mathbf{Z}_9 \times \mathbf{Z}_5 \cong \mathbf{Z}_2 \times \mathbf{Z}_2 \times \mathbf{Z}_{90} \quad \left\| \begin{array}{l} 2 \times 2 \times 2 \\ 3^2 \\ 5 \end{array} \right.$$

$$\mathbf{Z}_2 \times \mathbf{Z}_2 \times \mathbf{Z}_2 \times \mathbf{Z}_3 \times \mathbf{Z}_3 \times \mathbf{Z}_5 \cong \mathbf{Z}_2 \times \mathbf{Z}_6 \times \mathbf{Z}_{30} \quad \left\| \begin{array}{l} 2 \times 2 \times 2 \\ 3 \times 3 \\ 5 \end{array} \right.$$

Groups like $\mathbf{Z}_4 \times \mathbf{Z}_{90}$ are not of the form of those in Theorem 40, since 4 does not divide 90, but it isn't lost! In fact $\mathbf{Z}_4 \times \mathbf{Z}_{90}$ is the third group above, and $\mathbf{Z}_4 \times \mathbf{Z}_{90} \cong \mathbf{Z}_2 \times \mathbf{Z}_{180}$.

Theorem 39 and Theorem 40, separately or together, are often referred to as the **Fundamental theorem of finite abelian groups**.

## WHAT YOU SHOULD KNOW

■ What it means for two groups to be isomorphic.

■ How to determine whether two groups are isomorphic.

■ How to determine whether two groups are not isomorphic.

■ How to find all the finite abelian groups of a given order.

# EXERCISE 11

**1**   Determine whether or not $\mathbf{Z}_2 \times \mathbf{Z}_2 \cong \mathbf{Z}_4$. Justify your answer.

**2**   Determine whether or not $\mathbf{Z}_2 \times \mathbf{Z}_2 \cong V$. Justify your answer. (The group $V$ was introduced in Exercise 5, question 4.)

**3**   In the proof in Example 11.3.1 that $\mathbf{Z}_2 \times \mathbf{Z}_3 \cong \mathbf{Z}_6$, how many other isomorphisms could have been produced?

**4**   Let $G$ be a group, and let $g$ be a fixed element of $G$. Prove that the function $f : G \rightarrow G$ defined by $f(x) = g^{-1}xg$ is an isomorphism from $G$ onto itself.

**5**   Let $G$ and $H$ be groups, let $f : G \rightarrow H$ be an isomorphism and let $f(g) = h$. Prove that if $g$ has order $n$, then $h$ has order $n$.

**6**   Prove that the group $(\mathbf{R}, +)$ is isomorphic to the group $(\mathbf{R}^+, \times)$. [Hint: use the function $f(x) = e^x$.]

**7**   Prove that the groups $\mathbf{Z}$ and $3\mathbf{Z}$ under addition are isomorphic.

**8**   Prove that an abelian group cannot be isomorphic to a non-abelian group.

**9**   Prove that if $G$ and $H$ are groups then $G \times H \cong H \times G$.

**10**   Generalise the proof in Example 11.3.1 to prove that, if $m$ and $n$ are relatively prime, then $\mathbf{Z}_m \times \mathbf{Z}_n \cong \mathbf{Z}_{mn}$.

**11**   Let the positive integers $m$ and $n$ be relatively prime. Use the result of question 10, and in particular the surjectivity of the function $f : \mathbf{Z}_{mn} \rightarrow \mathbf{Z}_m \times \mathbf{Z}_n$ defined by $f([a]_{mn}) = ([a]_m, [a]_n)$, to prove that, for any integers $a$ and $b$, the equations $x \equiv a \pmod{m}$ and $x \equiv b \pmod{n}$ can be solved simultaneously.

*This result is usually called the Chinese remainder theorem.*

**12**   Prove that if $G \cong G'$ and that $H$ is a subgroup of $G$, then the image of $H$ under an isomorphism is a subgroup of $G'$.

**13**   Prove that if $G$ is cyclic, and $G \cong H$, then $H$ is cyclic.

*Exercises 14 to 19 are related in sequence and should be tackled together.*

**14**   Let $G$ be a group. Let $A$ denote the set of isomorphisms $G \rightarrow G$. Prove that $A$ with the operation of composition of functions is a group.

*The group $A$ of Exercise 14 is called the group of automorphisms, and written as $\mathrm{Aut}(G)$.*

**15**  Let $G$ be an abelian group and let $s$ be an integer. Prove that the function $f : A \rightarrow A$ defined by $f(x) = x^s$, satisfies $f(xy) = f(x)f(y)$.

**16**  Let $s \in \mathbf{Z}$. Define $f : \mathbf{Z} \rightarrow \mathbf{Z}$ by $f(a) = sa$, for all integers $a$. Prove that $f$ is an isomorphism if, and only if, $s = \pm 1$.

**17**  Let $s$ be an integer. Define $f : \mathbf{Z}_n \rightarrow \mathbf{Z}_n$ by $f([a]) = [sa]$, for $[a] \in \mathbf{Z}_n$. Prove that $f$ is an isomorphism if, and only if, $s$ and $n$ are relatively prime.

**18**  Prove that $Aut(\mathbf{Z}, +) \cong \mathbf{Z}_2$.

**19**  Let $U_n = \{[s] \in \mathbf{Z}_n : s \text{ and } n \text{ relatively prime}\}$. Then $U_n$ is a group under multiplication, and $Aut(\mathbf{Z}_n, +) \cong U_n$. (Hint: the fact that $U_n$ is a group is a generalisation of Theorem 14.)

**20**  Find all the abelian groups of order 18.

**21**  Find all the abelian groups of order 36.

**22**  Find all the abelian groups of order 180.

# 12

# *Permutations*

## 12.1 INTRODUCTION

Suppose that you have a set of four distinct objects. A permutation of these objects is a rearrangement of them among themselves. Thus if the objects are red, white, blue and green counters, then you could permute them by replacing the green counter by the blue one, and *vice versa*. Or you could replace blue by green, red by blue, and green by red. Both of these are examples of permutations. In practice, it is more convenient to label the four objects 1, 2, 3 and 4. Then the permutation which interchanges blue and green would be 'replace 1, 2, 3, 4 by 1, 2, 4, 3'. This permutation is called:

$$\begin{pmatrix} 1 & 2 & 3 & 4 \\ 1 & 2 & 4 & 3 \end{pmatrix}$$

to show that 4 replaces 3 and 3 replaces 4. Similarly the permutation which replaces 1, 2, 3, 4 by 3, 2, 4, 1 is called:

$$\begin{pmatrix} 1 & 2 & 3 & 4 \\ 3 & 2 & 4 & 1 \end{pmatrix}$$

showing that 3 replaces 1, 4 replaces 3 and 1 replaces 4.

Notice that the original order 1, 2, 3, 4 is not important. The permutation $\begin{pmatrix} 4 & 2 & 1 & 3 \\ 3 & 2 & 1 & 4 \end{pmatrix}$ means the same as $\begin{pmatrix} 1 & 2 & 3 & 4 \\ 1 & 2 & 4 & 3 \end{pmatrix}$; both of them show that 4 replaces 3 and 3 replaces 4. Similarly, $\begin{pmatrix} 1 & 2 & 3 & 4 \\ 3 & 2 & 4 & 1 \end{pmatrix}$, $\begin{pmatrix} 4 & 2 & 1 & 3 \\ 1 & 2 & 3 & 4 \end{pmatrix}$, $\begin{pmatrix} 2 & 3 & 4 & 1 \\ 2 & 4 & 1 & 3 \end{pmatrix}$ and $\begin{pmatrix} 2 & 1 & 4 & 3 \\ 2 & 3 & 1 & 4 \end{pmatrix}$ are identical permutations.

If the permutation $\begin{pmatrix} 1 & 2 & 3 & 4 \\ 1 & 2 & 4 & 3 \end{pmatrix}$ is followed by $\begin{pmatrix} 1 & 2 & 3 & 4 \\ 3 & 2 & 4 & 1 \end{pmatrix}$, then the effect is the permutation $\begin{pmatrix} 1 & 2 & 3 & 4 \\ 3 & 2 & 1 & 4 \end{pmatrix}$. It can be useful to consider the two permutations in the product 'above' one another, in the form

$$\begin{pmatrix} 1 & 2 & 3 & 4 \\ 1 & 2 & 4 & 3 \\ 3 & 2 & 1 & 4 \end{pmatrix}.$$

The first two rows show the effect of the first permutation, while the second two rows show the effect of the second permutation. The result of the first permutation followed by the second permutation is the permutation formed by taking the first and last rows, namely $\begin{pmatrix} 1 & 2 & 3 & 4 \\ 3 & 2 & 1 & 4 \end{pmatrix}$, showing that 3 replaces 1, and 1 replaces 3.

## 12.2 ANOTHER LOOK AT PERMUTATIONS

You can think of the ideas in Section 12.1 in a different way. Let $A$ be the set $\{1, 2, 3, 4\}$. Then the permutation $\begin{pmatrix} 1 & 2 & 3 & 4 \\ 1 & 2 & 4 & 3 \end{pmatrix}$ is really a bijection of the set $A$ to itself, defined in Fig. 12.1.

| 1 | $\rightarrow$ | 1 |    | 1 | $\rightarrow$ | 3 |
| 2 | $\rightarrow$ | 2 |    | 2 | $\rightarrow$ | 2 |
| 3 | $\rightarrow$ | 4 |    | 3 | $\rightarrow$ | 4 |
| 4 | $\rightarrow$ | 3 |    | 4 | $\rightarrow$ | 1 |

**Fig. 12.1**                          **Fig. 12.2**

In the same way, Fig. 12.2 shows the effect of the permutation $\begin{pmatrix} 1 & 2 & 3 & 4 \\ 3 & 2 & 4 & 1 \end{pmatrix}$. This leads to the following definition.

**Definition**: A **permutation** of a set $A$ is a bijection from $A$ to $A$.

*Looking at it this way, you can see that the method of combining permutations described in Section 12.1 is nothing more than the composition of functions.*

For example, returning to the case when $A = \{1,2,3,4\}$, for the permutation $\begin{pmatrix} 1 & 2 & 3 & 4 \\ 1 & 2 & 4 & 3 \end{pmatrix}$ followed by $\begin{pmatrix} 1 & 2 & 3 & 4 \\ 3 & 2 & 4 & 1 \end{pmatrix}$, you could draw the diagram in Fig. 12.3.

$$
\begin{array}{ccccc}
1 & \rightarrow & 1 & \rightarrow & 3 \\
2 & \rightarrow & 2 & \rightarrow & 2 \\
3 & \rightarrow & 4 & \rightarrow & 1 \\
4 & \rightarrow & 3 & \rightarrow & 4
\end{array}
$$

**Fig. 12.3**

This leads, as before, to the permutation $\begin{pmatrix} 1 & 2 & 3 & 4 \\ 3 & 2 & 1 & 4 \end{pmatrix}$.

*You may think that it is easier to think of a permutation as a function, and use the notation of Fig. 12.3, than to use the bracket notation. You may be right, but the bracket notation is traditional, and a little reflection should convince you that they are equivalent. Moreover, the bracket notation saves space on the page.*

It follows immediately from Theorem 34 that the set $S_A$ of permutations of a set $A$ under the operation of composition of functions is a group.

In the particular case when $A$ is the finite set $\{1, 2, \ldots, n\}$, the group of all permutations of $A$ is called the **symmetric group of degree $n$**, and is written $S_n$.

## 12.3 PRACTICE AT WORKING WITH PERMUTATIONS

### ■ *Example 12.3.1*

Calculate the result of the permutation $\begin{pmatrix} 1 & 2 & 3 \\ 2 & 3 & 1 \end{pmatrix}$ on three elements

followed by the permutation $\begin{pmatrix} 1 & 2 & 3 \\ 3 & 1 & 2 \end{pmatrix}$.

> *To do this you need to 'chase' each element through both permutations.*

In the first permutation $1 \to 2$, and in the second, $2 \to 1$, so combining the two permutations gives $1 \to 1$. Similarly, $2 \to 3 \to 2$, and $3 \to 1 \to 3$. So the combined permutation is $\begin{pmatrix} 1 & 2 & 3 \\ 1 & 2 & 3 \end{pmatrix}$. This is the identity permutation, in which each element is mapped to itself. Notice also that the permutations $\begin{pmatrix} 1 & 2 & 3 \\ 2 & 3 & 1 \end{pmatrix}$ and $\begin{pmatrix} 1 & 2 & 3 \\ 3 & 1 & 2 \end{pmatrix}$ are inverse.

> *The word 'inverse' is used in two senses here, and these two senses coincide. The two permutations, which are bijections, are inverse bijections. And each of them is also the inverse element of the other in the group sense of the word inverse.*

### ■ *Example 12.3.2*

Let $x = \begin{pmatrix} 1 & 2 & 3 & 4 & 5 \\ 3 & 1 & 5 & 4 & 2 \end{pmatrix}$ and $y = \begin{pmatrix} 1 & 2 & 3 & 4 & 5 \\ 2 & 3 & 4 & 5 & 1 \end{pmatrix}$.

Calculate $xy$ and $yx$.

> *Notice how the use of product notation has been slipped into permutations.*

As usual $xy$ means permutation $y$ followed by permutation $x$.

> *This notation is consistent with the notation for functions $f(g(a)) = (fg)(a)$.*

Chasing through the elements again, for the permutation $xy$, $1 \to 2 \to 1$, $2 \to 3 \to 5$, $3 \to 4 \to 4$, $4 \to 5 \to 2$, and $5 \to 1 \to 3$. The permutation $xy$ is given by

$$xy = \begin{pmatrix} 1 & 2 & 3 & 4 & 5 \\ 1 & 5 & 4 & 2 & 3 \end{pmatrix}.$$

Similarly, the permutation $yx = \begin{pmatrix} 1 & 2 & 3 & 4 & 5 \\ 4 & 2 & 1 & 5 & 3 \end{pmatrix}$.

*Notice that for permutations $xy$ is not generally equal to $yx$.*

## ■ Example 12.3.3

Let $x = \begin{pmatrix} 1 & 2 & 3 & 4 & 5 \\ 3 & 1 & 5 & 4 & 2 \end{pmatrix}$. Find the permutation $x^{-1}$.

You could do this in one of two ways. You might notice that $x^{-1} = \begin{pmatrix} 3 & 1 & 5 & 4 & 2 \\ 1 & 2 & 3 & 4 & 5 \end{pmatrix}$ and then rearrange the elements in the top row so that they come in the conventional order, at the same time moving the bottom row in a corresponding way. Thus $x^{-1} = \begin{pmatrix} 1 & 2 & 3 & 4 & 5 \\ 2 & 5 & 1 & 4 & 3 \end{pmatrix}$.

On the other hand, you could chase elements. Starting with 1, since $x$ takes $1 \to 3$, $x^{-1}$ takes $3 \to 1$. So $x^{-1}$ starts $x^{-1} = \begin{pmatrix} & & 3 & & \\ & & 1 & & \end{pmatrix}$. Then $x$ takes $2 \to 1$, so $x^{-1}$ takes $1 \to 2$. So $x^{-1}$ continues with $x^{-1} = \begin{pmatrix} 1 & & 3 & & \\ 2 & & 1 & & \end{pmatrix}$. And so on until $x^{-1} = \begin{pmatrix} 1 & 2 & 3 & 4 & 5 \\ 2 & 5 & 1 & 4 & 3 \end{pmatrix}$.

## ■ Example 12.3.4

Write out a multiplication table for $S_3$.

The possible permutations in $S_3$ are given by the six possible ways of ordering the numbers $a$, $b$ and $c$ as 1, 2 and 3 in the second row of the permutation $\begin{pmatrix} 1 & 2 & 3 \\ a & b & c \end{pmatrix}$.

They are therefore the six permutations.

$$e = \begin{pmatrix} 1 & 2 & 3 \\ 1 & 2 & 3 \end{pmatrix} \quad r = \begin{pmatrix} 1 & 2 & 3 \\ 2 & 3 & 1 \end{pmatrix} \quad r^2 = \begin{pmatrix} 1 & 2 & 3 \\ 3 & 1 & 2 \end{pmatrix}$$

$$x = \begin{pmatrix} 1 & 2 & 3 \\ 1 & 3 & 2 \end{pmatrix} \quad y = \begin{pmatrix} 1 & 2 & 3 \\ 3 & 2 & 1 \end{pmatrix} \quad z = \begin{pmatrix} 1 & 2 & 3 \\ 2 & 1 & 3 \end{pmatrix}$$

The group table for $S_3$ is shown in Fig. 12.4.

|       | $e$   | $r$   | $r^2$ | $x$   | $y$   | $z$   |
|-------|-------|-------|-------|-------|-------|-------|
| $e$   | $e$   | $r$   | $r^2$ | $x$   | $y$   | $z$   |
| $r$   | $r$   | $r^2$ | $e$   | $z$   | $x$   | $y$   |
| $r^2$ | $r^2$ | $e$   | $r$   | $y$   | $z$   | $x$   |
| $x$   | $x$   | $y$   | $z$   | $e$   | $r$   | $r^2$ |
| $y$   | $y$   | $z$   | $x$   | $r^2$ | $e$   | $r$   |
| $z$   | $z$   | $x$   | $y$   | $r$   | $r^2$ | $e$   |

**Fig. 12.4**

If you compare the group table in Fig. 12.4 with the table for $D_3$ in Example 5.2.2 you will see that they are the same, with

$$
\begin{array}{cccccc}
e & r & r^2 & x & y & z \\
\updownarrow & \updownarrow & \updownarrow & \updownarrow & \updownarrow & \updownarrow \\
I & R & S & X & Y & Z
\end{array}.
$$

This shows that the two groups $S_3$ and $D_3$ are isomorphic.

> *This method of showing that $S_3$ and $D_3$ are isomorphic is entirely valid, but does not generalise easily to other symmetry groups. Here is a more constructive way of showing the isomorphism.*

Consider $D_3$, the group of symmetries of an equilateral triangle T. Let $\phi \in D_3$. Then the image of each vertex of T is another vertex, and $\phi$ maps the set of vertices bijectively to itself. Label the vertices of T with the integers 1, 2 and 3 anticlockwise, and let $f(\phi)$ be the permutation

$$f(\phi) = \begin{pmatrix} 1 & 2 & 3 \\ \phi(1) & \phi(2) & \phi(3) \end{pmatrix}.$$

Then $f$ is a function $D_3 \to S_3$.

*Injection.* For $\phi$ and $\psi \in D_3$, if $f(\phi) = f(\psi)$, then

$$\begin{pmatrix} 1 & 2 & 3 \\ \phi(1) & \phi(2) & \phi(3) \end{pmatrix} = \begin{pmatrix} 1 & 2 & 3 \\ \psi(1) & \psi(2) & \psi(3) \end{pmatrix},$$

so $\phi(1) = \psi(1)$, $\phi(2) = \psi(2)$ and $\phi(3) = \psi(3)$. Therefore $\phi = \psi$, so $f$ is an injection.

As $D_3$ and $S_3$ both have 6 elements, it follows from Theorem 27 that $f$ is bijective.

Now, $f(\phi)$ can be written in the form $f(\phi) = \begin{pmatrix} a & b & c \\ \phi(a) & \phi(b) & \phi(c) \end{pmatrix}$

where $a$, $b$ and $c$ are the integers 1, 2 and 3 in any order. In particular, if $\psi$ is any element of $D_3$, $f(\phi) = \begin{pmatrix} \psi(1) & \psi(2) & \psi(3) \\ \phi(\psi(1)) & \phi(\psi(2)) & \phi(\psi(3)) \end{pmatrix}.$

Therefore

$$f(\phi)f(\psi) = \begin{pmatrix} \psi(1) & \psi(2) & \psi(3) \\ \phi(\psi(1)) & \phi(\psi(2)) & \phi(\psi(3)) \end{pmatrix} \begin{pmatrix} 1 & 2 & 3 \\ \psi(1) & \psi(2) & \psi(3) \end{pmatrix}$$

$$= \begin{pmatrix} 1 & 2 & 3 \\ \phi(\psi(1)) & \phi(\psi(2)) & \phi(\psi(3)) \end{pmatrix} = f(\phi\psi).$$

Therefore $f$ is an isomorphism.

## ■ *Example 12.3.5*

Find the order of the element $x = \begin{pmatrix} 1 & 2 & 3 & 4 & 5 & 6 \\ 3 & 1 & 2 & 4 & 6 & 5 \end{pmatrix}$ of $S_6$.

The order of an element $x$ is the smallest positive integer $k$ such that $x^k = e$. Considering in turn the powers of $x$ you find:

$$x^2 = \begin{pmatrix} 1 & 2 & 3 & 4 & 5 & 6 \\ 2 & 3 & 1 & 4 & 5 & 6 \end{pmatrix} \qquad x^3 = \begin{pmatrix} 1 & 2 & 3 & 4 & 5 & 6 \\ 1 & 2 & 3 & 4 & 6 & 5 \end{pmatrix}$$

$$x^4 = \begin{pmatrix} 1 & 2 & 3 & 4 & 5 & 6 \\ 3 & 1 & 2 & 4 & 5 & 6 \end{pmatrix} \qquad x^5 = \begin{pmatrix} 1 & 2 & 3 & 4 & 5 & 6 \\ 2 & 3 & 1 & 4 & 6 & 5 \end{pmatrix} \qquad x^6 = e.$$

Thus the order of $x = \begin{pmatrix} 1 & 2 & 3 & 4 & 5 & 6 \\ 3 & 1 & 2 & 4 & 6 & 5 \end{pmatrix}$ is 6.

## 12.4 EVEN AND ODD PERMUTATIONS

*The section which follows will show that permutations can be divided into two types, which will be called even and odd. There is a fair amount of preparatory work.*

Let $x$ be a permutation of $\{1, 2, \ldots, n\}$. Then consider the number of pairs of elements for which the order is reversed by $x$. That is, find $N(x) =$ the number of elements in the set $\{(i, j) : i < j \text{ and } x(i) > x(j)\}$.

For example, in $S_5$, let $x = \begin{pmatrix} 1 & 2 & 3 & 4 & 5 \\ 3 & 5 & 2 & 1 & 4 \end{pmatrix}$. Then since the pair $(1, 2)$ becomes $(3, 5)$, and $3 < 5$, the pair $(1, 2)$ is not reversed by $x$. On the other hand $(2, 3)$ becomes $(5, 2)$, and since $5 > 2$, the pair $(2, 3)$ is reversed by $x$. In fact the pairs $(1, 3)$, $(1, 4)$, $(2, 3)$, $(2, 4)$, $(2, 5)$ and $(3, 4)$ have all been reversed by $x$. This shows that $N(x) = 6$ or

$$N\left(\begin{pmatrix} 1 & 2 & 3 & 4 & 5 \\ 3 & 5 & 2 & 1 & 4 \end{pmatrix}\right) = 6.$$

Now consider the permutation $y = \begin{pmatrix} 1 & 2 & 3 & 4 & 5 \\ 3 & 1 & 5 & 4 & 2 \end{pmatrix}$. In this case, the pairs $(1, 2)$, $(1, 5)$, $(3, 4)$, $(3, 5)$ and $(4, 5)$ have been reversed, showing that $N(y) = 5$.

Finally consider the product $yx = \begin{pmatrix} 1 & 2 & 3 & 4 & 5 \\ 5 & 2 & 1 & 3 & 4 \end{pmatrix}$. The reversed pairs

are $(1,2)$, $(1,3)$, $(1,4)$, $(1,5)$ and $(2,3)$ and $N(yx) = 5$.

Notice that $N(x)$ is even, $N(y)$ is odd and $N(yx)$ is odd.

*This is an example of a general result about permutations, which can be summarised informally by the table in Fig. 12.5.*

| $N(x)$ | $N(y)$ | $N(yx)$ |
|--------|--------|---------|
| even | even | even |
| even | odd | odd |
| odd | even | odd |
| odd | odd | even |

**Fig. 12.5**

Here is a proof of the result shown informally in the table.

**Theorem 41:** For any pair of permutations $x$ and $y$ of $S_n$, $(-1)^{N(yx)} = (-1)^{N(y)+N(x)}$.

**Proof**: Suppose that there are $M$ pairs overall.

Consider the pairs which are reversed by $x$. There are $N(x)$ of them. When $y$ acts on the results of $x$, some of them, $k$ say, will be reversed back, while the others, $N(x) - k$ of them, will remain reversed.

Now consider the remainder of the original $M$ pairs which are not reversed by $x$. There are $M - N(x)$ of these. When $y$ acts on the results of $x$, $N(y) - k$ of them will become reversed, and the remainder, $M - N(x) - (N(y) - k)$ of them, will remain unreversed.

Counting the contributions to the total of the reverses under $yx$ gives $N(x) - k$ of the first type, and $N(y) - k$ of the second. The total, $N(yx)$, is therefore given by $N(yx) = N(x) + N(y) - 2k$. Thus, $(-1)^{N(yx)} = (-1)^{N(y)+N(x)}$. ∎

This justifies the results given in Fig. 12.5.

**Definition**: Suppose that $x \in S_n$, the group of permutations on $n$ symbols. Then $x$ is an **even permutation** if $N(x)$ is even, and $x$ is an **odd permutation** if $N(x)$ is odd.

Suppose that you have a permutation $x = \begin{pmatrix} 1 & 2 & \ldots & n \\ x(1) & x(2) & \ldots & x(n) \end{pmatrix}$.

Then if you draw a line from each number in the top row to where that number appears in the bottom row, it will turn out that $N(x) =$ the number of intersections.

For example, consider the permutation $x = \begin{pmatrix} 1 & 2 & 3 & 4 \\ 2 & 3 & 4 & 1 \end{pmatrix}$. Then Fig. 12.6 shows the lines with three intersections. The claim is that $N(x) = 3$.

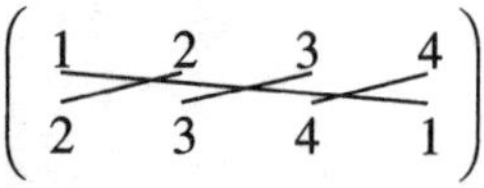

**Fig. 12.6**

Theorem 42 justifies this claim.

**Theorem 42:** Let $x = \begin{pmatrix} 1 & 2 & \ldots & n \\ x(1) & x(2) & \ldots & x(n) \end{pmatrix}$ be a permutation on $n$ symbols. Then $N(x) =$ the number of intersections of lines drawn from each number in the top row to where that number appears in the bottom row.

**Proof**: The elements of the first row are exactly the same as the elements of the second row. But in the first row, the elements are in numerical order.

Suppose that $i < j$. Then $x(i)$ will appear to the left of $x(j)$ in the second row of the permutation $x$.

Now suppose first that $x(i) < x(j)$. Then $x(i)$ will also appear to the left of $x(j)$ in the first row of the permutation $x$, as in Fig. 12.7.

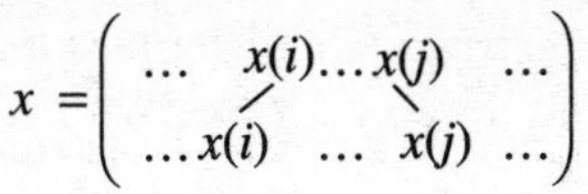

$$x = \begin{pmatrix} \ldots & x(i) \ldots x(j) & \ldots \\ \ldots x(i) & \ldots & x(j) \ldots \end{pmatrix} \qquad x = \begin{pmatrix} \ldots & x(j) \ldots x(i) & \ldots \\ \ldots & x(i) \ldots x(j) & \ldots \end{pmatrix}$$

**Fig. 12.7** 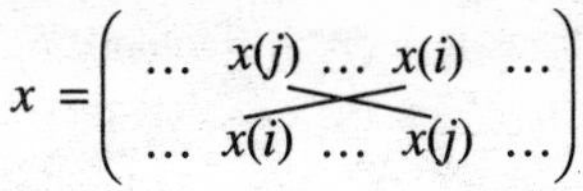                     **Fig. 12.8**

Now suppose that $x(i) > x(j)$. Then $x(i)$ will appear to the right of $x(j)$ in the first row of the permutation $x$, as in Fig. 12.8.

There are $n$ lines, one for each $i$, or, if you prefer, one for each $x(i)$, and each intersection of two of these lines corresponds to a reversal $i < j : x(i) > x(j)$.

More precisely, there is a one-to-one correspondence between the set of intersections of lines and the set of pairs $\{(i, j) : i < j \text{ and } x(i) > x(j)\}$.

Therefore $N(x) =$ the number of intersections. ∎

### ■ *Example 12.4.1*

For the permutation $x = \begin{pmatrix} 1 & 2 & 3 & 4 & 5 \\ 3 & 5 & 2 & 1 & 4 \end{pmatrix}$, six of the 20 pairs of lines

are intersecting, and 14 are non-intersecting.

Figure 12.9 shows an intersecting pair and a non-intersecting pair and Fig. 12.10 shows all the intersections.

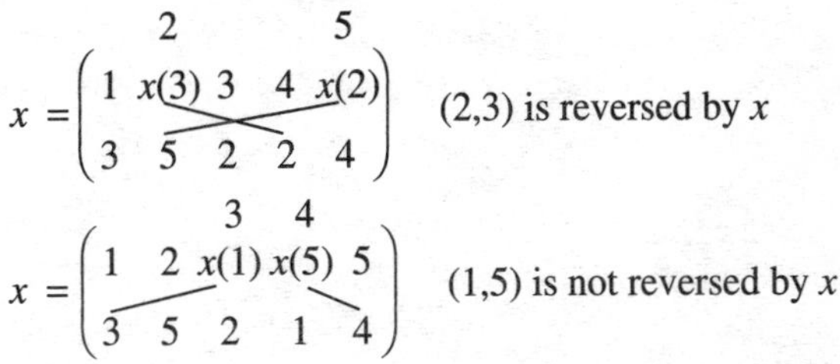

$$x = \begin{pmatrix} 1 & x(3) & 3 & 4 & x(2) \\ 3 & 5 & 2 & 2 & 4 \end{pmatrix} \qquad (2,3) \text{ is reversed by } x$$

$$x = \begin{pmatrix} 1 & 2 & x(1) & x(5) & 5 \\ 3 & 5 & 2 & 1 & 4 \end{pmatrix} \qquad (1,5) \text{ is not reversed by } x$$

**Fig. 12.9**

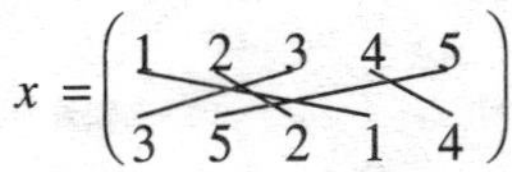

$$x = \begin{pmatrix} 1 & 2 & 3 & 4 & 5 \\ 3 & 5 & 2 & 1 & 4 \end{pmatrix}$$

**Fig. 12.10**

It follows that $N(x) = 6$, and $x$ is an even permutation.

Thus the set of permutations is divided into even permutations and odd permutations.

> *This theme will be put on one side for the moment, while a new notation for writing permutations is introduced.*

## 12.5 CYCLES

Cycle notation is an alternative, shorter notation, which is often used for permutations of finite sets.

The permutation $\begin{pmatrix} a_1 & a_2 & a_3 & \dots & a_n \\ a_2 & a_3 & a_4 & \dots & a_1 \end{pmatrix} \in S_n$ is a **cycle of length** $n$.

This is shown in Fig. 12.11, where each number is mapped by the permutation into the number one place round the circle anti-clockwise.

The cycle $\begin{pmatrix} a_1 & a_2 & a_3 & \dots & a_n \\ a_2 & a_3 & a_4 & \dots & a_1 \end{pmatrix} \in S_n$ is written $\left( a_1 a_2 a_3 \dots a_n \right)$ which

means, reading from the left, that $a_1 \to a_2$, $a_2 \to a_3$, $a_3 \to a_4$, and so on, finishing with the understanding that $a_n \to a_1$.

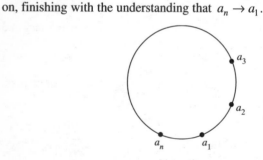

**Fig. 12.11**

> *Notice that if you talk about the cycle (12478635), there is no indication whether this comes from $S_8$, or possibly $S_9$, $S_{29}$, or $S_n$ for any other $n \geq 8$. If this is likely to cause confusion, then you have to make the underlying set clear. Any number not mentioned in the cycle remains unchanged.*

## ■ *Example 12.5.1*

Consider $(124) \in S_6$ so $(124) = \begin{pmatrix} 1 & 2 & 3 & 4 & 5 & 6 \\ 2 & 4 & 3 & 1 & 5 & 6 \end{pmatrix}$. Consider also $(1256) \in S_6$.

Then the product $(1256)(124) = \begin{pmatrix} 1 & 2 & 3 & 4 & 5 & 6 \\ 5 & 4 & 3 & 2 & 6 & 1 \end{pmatrix}$. Similarly $(124)(1256) = \begin{pmatrix} 1 & 2 & 3 & 4 & 5 & 6 \\ 4 & 5 & 3 & 1 & 6 & 2 \end{pmatrix}$. Notice that these two results are different, so the cycles $(124)$ and $(1256)$ do not commute.

## ■ *Example 12.5.2*

Consider $(14) \in S_6$ and $(235) \in S_6$. Then the permutations $(14)(235)$ and $(235)(14)$ are given by $(14)(235) = \begin{pmatrix} 1 & 2 & 3 & 4 & 5 & 6 \\ 4 & 3 & 5 & 1 & 2 & 6 \end{pmatrix}$ and $(235)(14) = \begin{pmatrix} 1 & 2 & 3 & 4 & 5 & 6 \\ 4 & 3 & 5 & 1 & 2 & 6 \end{pmatrix}$.

The difference between Example 12.5.1 and Example 12.5.2 is that in Example 12.5.2 the cycles $(14)$ and $(235)$ have no numbers in common, whereas $(124)$ and $(1256)$ have numbers in common.

In general, the cycles in a collection of cycles are said to be **disjoint** if no number in them is repeated.

Disjoint cycles always commute with each other because they operate on disjoint subsets of the set $\{1, 2, 3, \ldots, n\}$.

Here is an example to show you how to write a permutation as the product of disjoint cycles.

To write the permutation $\begin{pmatrix} 1 & 2 & 3 & 4 & 5 & 6 & 7 & 8 \\ 2 & 4 & 5 & 1 & 7 & 8 & 3 & 6 \end{pmatrix} \in S_8$ as a product of disjoint cycles, start with the 1, and follow through what happens to 1 and its successive images; you find that $1 \to 2 \to 4 \to 1$. So start by writing $(124)$. Then take a number not already used, say 3, and follow that through. You find that $3 \rightarrow 5 \rightarrow 7 \rightarrow 3$, and write $(357)$. Take a number not already used, say 6, and you find that $6 \rightarrow 8 \rightarrow 6$, leading to $(68)$. All the numbers are now accounted for. Therefore

$$\begin{pmatrix} 1 & 2 & 3 & 4 & 5 & 6 & 7 & 8 \\ 2 & 4 & 5 & 1 & 7 & 8 & 3 & 6 \end{pmatrix} = (124)(357)(68),$$ which is a product of

disjoint cycles.

For the permutation $\begin{pmatrix} 1 & 2 & 3 & 4 & 5 & 6 & 7 & 8 \\ 2 & 4 & 5 & 1 & 7 & 6 & 8 & 3 \end{pmatrix}$, the corresponding

result would have been $\begin{pmatrix} 1 & 2 & 3 & 4 & 5 & 6 & 7 & 8 \\ 2 & 4 & 5 & 1 & 7 & 6 & 8 & 3 \end{pmatrix} = (124)(3578).$

Notice the importance of the numbers in the cycles being disjoint.

These two examples show the way to a general result.

**Theorem 43:** Any permutation of a finite set can be written as a product of disjoint cycles. The product is unique up to the order in which the cycles appear, and the different ways of writing each cycle.

*This result will not be proved formally. The proof, which is rather tedious, essentially follows the method given in the example above.* ∎

## ■ *Example 12.5.3*

Express $x = \begin{pmatrix} 1 & 2 & 3 & 4 & 5 & 6 & 7 & 8 \\ 1 & 3 & 5 & 7 & 2 & 4 & 6 & 8 \end{pmatrix}$, $x^2$ and $x^{-1}$ as the product of

disjoint cycles.

Using the method of this section, $x = (476)(235)$. For $x^2$, each number is moved by two positions in the cycle for $x$, so $x^2 = (467)(253)$. For $x^{-1}$, each cycle in $x$ must be reversed, so $x^{-1} = (467)(253)$.

Notice that $x^2 = x^{-1}$, and $x^3$ is the identity permutation.

## 12.6 Transpositions

A **transposition** is a cycle of length 2. Examples of transpositions are (34) and (26).

Notice that $(a_1 a_2 a_3 \ldots a_n) = (a_1 a_n)(a_1 a_{n-1}) \ldots (a_1 a_3)(a_1 a_2)$, so every cycle can be written as a product of transpositions.

> *The easiest way to see this is simply to multiply out the right-hand side and see the pattern of what is happening. If you want a formal proof of the result you would need to use the principle of mathematical induction.*

It follows that every permutation can be written as the product of transpositions. First write the permutation as the product of cycles, and then as the product of transpositions.

To illustrate this, in Example 12.5.3, $x = (476)(235)$; as products of transpositions, $(476) = (46)(47)$ and $(235) = (25)(23)$; so $x = (46)(47)(25)(23)$.

Notice that this is not a unique representation. For example, the cycle (476) is also equal to (74)(76). So $x$ can also be written as $x = (74)(76)(25)(23)$.

Indeed the representation of a permutation as a product of transpositions can never be unique, because $(ab)(ba)$ can be inserted at any point into any product, and also, for any transposition $(ab)$, $(ab) = (1a)(1b)(1a)$.

### ■ *Example 12.6.1*

Express $(12)(476)(1235)$ as the product of transpositions.

> *There are two distinct ways you could set about this. You could ignore the fact that the original permutation is not the product of disjoint cycles, and simply use the result at the beginning of this section directly; or you could simplify the permutation as the product of disjoint cycles, and then write it as the product of transpositions.*

Using the first method, you get $(12)(46)(47)(15)(13)(12)$.

Using the second method, $(12)(476)(1235) = (476)(235)$, leading to $(12)(476)(1235) = (46)(47)(25)(23)$.

## ■ *Example 12.6.2*

Write down the inverse of the permutation $(12)(46)(47)(23)(45)$.

The inverse of a transposition is itself, so $(12)^{-1} = (12)$. Using the fact that $(xy)^{-1} = y^{-1}x^{-1}$, the inverse of $(12)(46)(47)(23)(45)$ is $(45)^{-1}(23)^{-1}(47)^{-1}(46)^{-1}(12)^{-1}$, which is $(45)(23)(47)(46)(12)$.

Although the number of transpositions required to write a given permutation is not unique, what is true is that each permutation, when written as a product of transpositions, will always give an even number of transpositions, or will always give an odd number of transpositions. This will be proved in Theorem 45 below.

It is probably not a surprise to find that transpositions are odd permutations.

**Theorem 44**: Transpositions are odd permutations.

**Proof**: Let the transposition $t = (i\,j) \in S_n$, where $i < j$. Then $t$ interchanges $i$ and $j$ and is illustrated in Fig. 12.12.

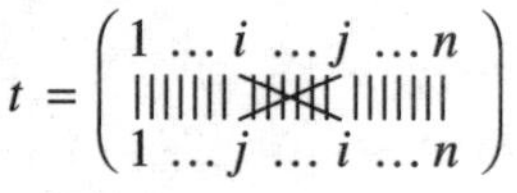

$$t = \begin{pmatrix} 1 \dots i \dots j \dots n \\ |||||||\,\mathrm{X\!X\!X}\,||||||| \\ 1 \dots j \dots i \dots n \end{pmatrix}$$

**Fig. 12.12**

The number of intersections is $2(j - i - 1) + 1$, which is odd. ■

**Theorem 45**: When an even permutation is written as a product of transpositions, the number of transpositions is even, and when an odd permutation is written as a product of transpositions, the number of transpositions is odd.

**Proof**: From Theorem 44, for every transposition $t$, $(-1)^{N(t)} = -1$. Let $x \in S_n$ be any permutation, and suppose that $x$ is written as the product of $k$ transpositions $x = t_k t_{k-1} \dots t_2 t_1$.

Then, from an obvious extension of Theorem 41,

$$(-1)^{N(x)} = (-1)^{N(t_k)+N(t_{k-1})+...+N(t_2)+N(t_1)}$$

$$= (-1)^{N(t_k)} \times (-1)^{N(t_{k-1})} \times ... \times (-1)^{N(t_2)} \times (-1)^{N(t_1)}$$

$$= \overbrace{(-1)(-1)...(-1)(-1)}^{k \text{ times}}$$

$$= (-1)^k$$

Therefore, if $N(x)$ is even, $k$ is even, and if $N(x)$ is odd, $k$ is odd. ∎

## 12.7 THE ALTERNATING GROUP

You may now be guessing that the set $A_n$ of even permutations of $S_n$ is a subgroup of $S_n$. To prove this, suppose that $x \in A_n$ and $y \in A_n$.

Since $x$ and $y$ can be written as a product of an even number of transpositions, so can $xy$, because even + even = even. Thus if $x \in A_n$ and $y \in A_n$ then $xy \in A_n$.

Notice that the identity permutation $e$ involves no reversals, that is, $N(e) = 0$, and is therefore even.

If $x$ reverses the order of $i, j$, then $x^{-1}$ reverses the order of $x(i), x(j)$. So the number of pairs reversed by $x$ is equal to the number of pairs reversed by $x^{-1}$; that is $N(x) = N\left(x^{-1}\right)$. So, if $x \in A_n$, then $x^{-1} \in A_n$.

By Theorem 21, this establishes that $A_n$ is a subgroup of $S_n$.

**Definition**: The subgroup $A_n$ of the group $S_n$ of permutations on $n$ symbols is called the **alternating group on $n$ symbols**.

It should not be a surprise that half the permutations are even and half are odd, but to prove it you need to set up a bijection between the even permutations and the odd permutations.

An example of such a bijection is $f : A_n \to S_n - A_n$ defined by $f(x) = (12)x$. It is left as an exercise, Question 15 in Exercises 12, to prove that $f$ is a bijection.

Thus $A_3$ has three elements, and $A_4$ has twelve elements. In general $A_n$ has $\frac{1}{2}n!$ elements.

For future reference, Fig 12.13 shows the alternating group $A_4$.

|       | $e$   | $a$   | $b$   | $c$   | $x$   | $y$   | $z$   | $t$   | $x^2$ | $y^2$ | $z^2$ | $t^2$ |
|-------|-------|-------|-------|-------|-------|-------|-------|-------|-------|-------|-------|-------|
| $e$   | $e$   | $a$   | $b$   | $c$   | $x$   | $y$   | $z$   | $t$   | $x^2$ | $y^2$ | $z^2$ | $t^2$ |
| $a$   | $a$   | $e$   | $c$   | $b$   | $z$   | $t$   | $x$   | $y$   | $t^2$ | $z^2$ | $y^2$ | $x^2$ |
| $b$   | $b$   | $c$   | $e$   | $a$   | $t$   | $z$   | $y$   | $x$   | $y^2$ | $x^2$ | $t^2$ | $z^2$ |
| $c$   | $c$   | $b$   | $a$   | $e$   | $y$   | $x$   | $t$   | $z$   | $z^2$ | $t^2$ | $x^2$ | $y^2$ |
| $x$   | $x$   | $t$   | $y$   | $z$   | $x^2$ | $t^2$ | $y^2$ | $z^2$ | $e$   | $c$   | $a$   | $b$   |
| $y$   | $y$   | $z$   | $x$   | $t$   | $z^2$ | $y^2$ | $t^2$ | $x^2$ | $c$   | $e$   | $b$   | $a$   |
| $z$   | $z$   | $y$   | $t$   | $x$   | $t^2$ | $x^2$ | $z^2$ | $y^2$ | $a$   | $b$   | $e$   | $c$   |
| $t$   | $t$   | $x$   | $z$   | $y$   | $y^2$ | $z^2$ | $x^2$ | $t^2$ | $b$   | $a$   | $c$   | $e$   |
| $x^2$ | $x^2$ | $z^2$ | $t^2$ | $y^2$ | $e$   | $b$   | $c$   | $a$   | $x$   | $z$   | $t$   | $y$   |
| $y^2$ | $y^2$ | $t^2$ | $z^2$ | $x^2$ | $b$   | $e$   | $a$   | $c$   | $t$   | $y$   | $x$   | $z$   |
| $z^2$ | $z^2$ | $x^2$ | $y^2$ | $t^2$ | $c$   | $a$   | $e$   | $b$   | $y$   | $t$   | $z$   | $x$   |
| $t^2$ | $t^2$ | $y^2$ | $x^2$ | $z^2$ | $a$   | $c$   | $b$   | $e$   | $z$   | $x$   | $y$   | $t$   |

**Fig. 12.13** *The alternating group* $A_4$

In Fig. 12.13, the elements $a = (12)(34)$, $b = (13)(24)$ and $c = (14)(23)$ are each of order 2. The elements $x = (234)$, $y = (143)$, $z = (124)$ and $t = (132)$ each have order 3.

## WHAT YOU SHOULD KNOW

■ What a permutation of a set $S$ is.

■ The notation for permutations of a finite set, and how to combine them.

■ That the permutations of a set form a group.

■ The meaning of 'even' and 'odd' as applied to permutations.

■ How to determine whether a given permutation is even or odd.

- What cycle notation is, and how to use it.

- The meaning of 'transposition', and how to write every permutation as a product of transpositions.

- Each permutation, when written as a product of transpositions, will always give an even number of transpositions, or will always give an odd number of transpositions.

- The meaning of 'alternating group'.

## EXERCISE 12

**1** The permutations $a$, $b$ and $c$ are taken from $S_5$.

$$a = \begin{pmatrix} 1 & 2 & 3 & 4 & 5 \\ 5 & 3 & 4 & 1 & 2 \end{pmatrix} \quad b = \begin{pmatrix} 1 & 2 & 3 & 4 & 5 \\ 2 & 3 & 4 & 5 & 1 \end{pmatrix} \quad c = \begin{pmatrix} 1 & 2 & 3 & 4 & 5 \\ 5 & 3 & 2 & 4 & 1 \end{pmatrix}$$

Calculate the permutations: $ab$, $ba$, $a^2b$, $ac^{-1}$, $(ac)^{-1}$, $c^{-1}ac$.

**2** Using the permutations from question 1, solve for $x$ the equations $ax = b$ and $axb = c$.

**3** Using the permutations from question 1, find the number of reversals in each of $a$, $b$ and $c$.

**4** Let

$$a = \begin{pmatrix} 1 & 2 & 3 & 4 & 5 \\ 2 & 4 & 1 & 3 & 5 \end{pmatrix} \quad b = \begin{pmatrix} 1 & 2 & 3 & 4 & 5 \\ 4 & 2 & 1 & 5 & 3 \end{pmatrix} \quad c = \begin{pmatrix} 1 & 2 & 3 & 4 & 5 \\ 1 & 4 & 2 & 3 & 5 \end{pmatrix}.$$

Find the number of reversals in each of $a$, $b$ and $c$, and deduce whether the number of reversals in each of $ab$, $bc$ and $ca$ is even or odd.

**5** Use the method of lines drawn from the top row of a permutation to the bottom row, as in Theorem 42, to determine whether the permutations $\begin{pmatrix} 1 & 2 & 3 & 4 & 5 \\ 1 & 3 & 5 & 2 & 4 \end{pmatrix}$ and $\begin{pmatrix} 1 & 2 & 3 & 4 & 5 \\ 3 & 1 & 4 & 2 & 5 \end{pmatrix}$ are even or odd.

**6** Write each of the following permutations in cycle notation, as products of disjoint cycles.

$$\begin{pmatrix} 1 & 2 & 3 & 4 & 5 \\ 3 & 4 & 5 & 2 & 1 \end{pmatrix} \quad \begin{pmatrix} 1 & 2 & 3 & 4 & 5 \\ 3 & 1 & 4 & 2 & 5 \end{pmatrix} \quad \begin{pmatrix} 1 & 2 & 3 & 4 & 5 \\ 4 & 5 & 3 & 1 & 2 \end{pmatrix}.$$

**7**   Write each of the following products of disjoint cycles in $S_6$ in the notation used in question 1.

(1)   (123)(46)

(2)   (12346)

(3)   (12)(346)

**8**   Write each of the following permutations in $S_7$ as the product of disjoint cycles.

(1)   (12)(347)(132)

(2)   (56)(347)(45)(132)

(3)   (34)(143)(43)(132)

(4)   (16)(15)(14)(13)(12)

**9**   In question 4 of Exercise 5, you were asked to draw up a group table for the symmetries of a rectangle which was not a square. Labelling the vertices of the rectangle with the integers 1, 2, 3, 4 and working as in Example 12.3.4, show that this group is isomorphic to the subgroup $\{e, (12)(34), (13)(24), (14)(23)\}$ of $S_4$. Write out the table for this group of permutations.

**10**   The group $S_4$ has 24 elements and $S_3$ has six elements. Investigate whether $S_4 \cong Z_4 \times S_3$ or whether $S_4 \cong Z_2 \times Z_2 \times S_3$.

**11**   Prove that the order of a cycle is equal to its length.

**12**   Prove that the order of any permutation is equal to the least common multiple of the lengths of its component disjoint cycles. (Notice that this gives a quicker method for finding the order of the element $x$ in Example 12.3.5. The element $x = (132)(56)$ and so has order $3 \times 2 = 6$.)

**13**   Prove that every element of $S_n$ can be written as the product of the transpositions $(1\,2)$, $(1\,3)$, ... , $(1\,n)$.

**14**   Prove that every element of $S_n$ can be written as the product of $(1\,2\ldots n-1)$ and $(n-1\ \ n)$.

**15**   Define $f : A_n \to S_n - A_n$ by $f(x) = (12)x$ for $x \in A_n$. Prove that $f$ is a bijection. Hence show that the number of elements in $A_n$ is $\frac{1}{2}n!$.

# 13

# *Dihedral groups*

## 13.1 INTRODUCTION

In Chapter 5 when you first met the idea of a group, you studied $D_3$, the group of symmetries of an equilateral triangle.

Then, in question 5 of Exercise 10, you were asked to show that for any subset $X$ of $A$, the set $\{f \in$ set of bijections $A \to A : f(X) = X\}$ is a subgroup of the group of bijections $A \to A$.

> *Recall that $f(X) = X$ is not the same as saying that $f(x) = x$ for all $x \in X$. For example, if $A = \{1,2,3\}$, $X = \{1,2\}$, and $f$ is the transposition $(12)$, then $f(X) = X$ but $f(1) \neq 1$ and $f(2) \neq 2$.*

**Definition:** Let $A$ be the real plane. Then for any subset $X$ of $A$, a distance-preserving function $f : A \to A$ such that $f(X) = X$ is called a **symmetry** of $X$. The symmetries of $X$ form a subgroup of the group of bijections $A \to A$; this is called the **group of symmetries of $X$**.

> *A distance-preserving function has the property that, for every pair of points $P$ and $Q$ in the plane, the distance between $f(P)$ and $f(Q)$ is the same as the distance between $P$ and $Q$.*

> *Every distance-preserving function $A \to A$ from the plane to*
> *itself is actually a bijection, and the set of distance-preserving*
> *functions $A \to A$ forms a subgroup of the group of bijections.*
> *Neither of these statements will be proved here.*

**Definition:** The group of symmetries of a regular $n$-sided polygon is
called **the dihedral group** $D_n$.

The group $D_3$ of symmetries of an equilateral triangle, is an example of
a dihedral group.

> *The word 'dihedral' means two-faced. This counts the front of the*
> *regular polygon as one face, and the back as the other.*

If X is any polygon and $\phi$ is a symmetry of X, then the image of each
vertex of X is another vertex and, since $\phi$ is a bijection, $\phi$ maps the
set of vertices bijectively into itself. If X has $n$ vertices $1, 2, \dots, n$ and
$f(\phi)$ is the permutation $f(\phi) = \begin{pmatrix} 1 & 2 & \dots & n \\ \phi(1) & \phi(2) & \dots & \phi(n) \end{pmatrix}$, then $f$ is a
function from the symmetries of X to $S_n$. As in Example 12.3.4, this
function is injective and has the property that, for any two symmetries
$\phi$ and $\psi$ of X, $f(\phi\psi) = f(\phi)f(\psi)$.

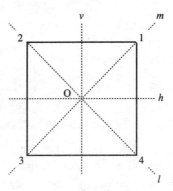

**Fig. 13.1**

Consider $D_4$, the group of symmetries of a square. The symmetries of
the square in Fig. 13.1 are $H$, $V$, $L$, $M$, $R$, $R^2$, $R^3$ and $I$, where
● $H$ means  'Reflect in the axis $h$.'

- $V$ means    'Reflect in the axis $v$.'
- $L$ means    'Reflect in the axis $l$.'
- $M$ means    'Reflect in the axis $m$.'
- $R$ means    'Rotate by 90° anticlockwise about O.'
- $R^2$ means  'Rotate by 180° anticlockwise about O.'
- $R^3$ means  'Rotate by 270° anticlockwise about O.'
- $I$ means    'Do nothing.'

The images of these symmetries under the function $f : D_4 \to S_4$ described above are $f(H) = (14)(23)$, $f(V) = (12)(34)$, $f(L) = (13)$, $f(M) = (24)$, $f(R) = (1234)$, $f(R^2) = (13)(24)$, $f(R^3) = (4321)$ and $f(I) = e$. These eight permutations form a subgroup of $S_4$, and, as in Example 12.3.4, $D_4$ is isomorphic to this subgroup of $S_4$.

Fig. 13.2 shows the table for $D_4$.

> To work out the entries quickly, you may find it useful to make a square out of card and write 1, 2, 3, 4 on both faces. Alternatively, you can multiply the permutations: for example, $f(H)f(R) = (14)(23)(1234) = (13) = f(L)$, so $HR = L$.

|       | $I$   | $R$   | $R^2$ | $R^3$ | $H$   | $L$   | $V$   | $M$   |
|-------|-------|-------|-------|-------|-------|-------|-------|-------|
| $I$   | $I$   | $R$   | $R^2$ | $R^3$ | $H$   | $L$   | $V$   | $M$   |
| $R$   | $R$   | $R^2$ | $R^3$ | $I$   | $M$   | $H$   | $L$   | $V$   |
| $R^2$ | $R^2$ | $R^3$ | $I$   | $R$   | $V$   | $M$   | $H$   | $L$   |
| $R^3$ | $R^3$ | $I$   | $R$   | $R^2$ | $L$   | $V$   | $M$   | $H$   |
| $H$   | $H$   | $L$   | $V$   | $M$   | $I$   | $R$   | $R^2$ | $R^3$ |
| $L$   | $L$   | $V$   | $M$   | $H$   | $R^3$ | $I$   | $R$   | $R^2$ |
| $V$   | $V$   | $M$   | $H$   | $L$   | $R^2$ | $R^3$ | $I$   | $R$   |
| $M$   | $M$   | $H$   | $L$   | $V$   | $R$   | $R^2$ | $R^3$ | $I$   |

**Fig 13.2** *The dihedral group $D_4$*

You can see various patterns in this table. For example, the rotations are all in the top left and bottom right in the results, and the reflections are in the opposite corners. However, this kind of table does not bring out the full structure of the dihedral group.

## 13.2 TOWARDS A GENERAL NOTATION

Working out the entries in a dihedral group using this notation is time-consuming. It is better to work with one rotation, and one reflection.

Suppose that $a$ is a rotation through 90° anti-clockwise, and that $b$ is any reflection. Then $a^4 = b^2 = e$. You can also check from the table in Fig. 13.2 that, no matter which reflection $b$ you choose, $aba = b$.

In terms of $a$ and $b$ the table of Fig. 13.2 now becomes Fig. 13.3.

|        | $e$    | $a$    | $a^2$  | $a^3$  | $b$    | $ba$   | $ba^2$ | $ba^3$ |
|--------|--------|--------|--------|--------|--------|--------|--------|--------|
| $e$    | $e$    | $a$    | $a^2$  | $a^3$  | $b$    | $ba$   | $ba^2$ | $ba^3$ |
| $a$    | $a$    | $a^2$  | $a^3$  | $e$    | $ba^3$ | $b$    | $ba$   | $ba^2$ |
| $a^2$  | $a^2$  | $a^3$  | $e$    | $a$    | $ba^2$ | $ba^3$ | $b$    | $ba$   |
| $a^3$  | $a^3$  | $e$    | $a$    | $a^2$  | $ba$   | $ba^2$ | $ba^3$ | $b$    |
| $b$    | $b$    | $ba$   | $ba^2$ | $ba^3$ | $e$    | $a$    | $a^2$  | $a^3$  |
| $ba$   | $ba$   | $ba^2$ | $ba^3$ | $b$    | $a^3$  | $e$    | $a$    | $a^2$  |
| $ba^2$ | $ba^2$ | $ba^3$ | $b$    | $ba$   | $a^2$  | $a^3$  | $e$    | $a$    |
| $ba^3$ | $ba^3$ | $b$    | $ba$   | $ba^2$ | $a$    | $a^2$  | $a^3$  | $e$    |

**Fig. 13.3** *The dihedral group* $D_4$

The relations $a^4 = b^2 = e$ and $aba = b$, enable you to carry out any calculation in $D_4$ without the table. Remember that, in $D_4$, $a^{-1} = a^3$, so that from $aba = b$ you can deduce that $ab = ba^3$ and $ba = a^3b$.

For example, if you wish to calculate $ba^2 \circ ba$, you say
$ba^2 \circ ba = ba(aba) = bab = b(ab) = b(ba^3) = a^3$.

Similarly $ba \circ ba^2 = baba^2 = b(aba)a = b^2a = a$.

The group is said to be 'generated' by $a$ and $b$, together with the relations $a^4 = b^2 = e$ and $aba = b$.

## ■ Example 13.2.1

Use the relations $a^4 = b^2 = e$ and $ab = ba^3$ to prove that $a^2b = ba^2$ and $a^3b = ba$.

> *In the solution to this example, notice how the use of the relation enables the b to pass through the as. The solutions are written out in more detail than you might need to write them. The second solution uses the first result.*

$$a^2b = a(ab) = a(ba^3) = (ab)a^3 = (ba^3)a^3 = b(a^3a^3) = ba^2$$
$$a^3b = a(a^2b) = a(ba^2) = (ab)a^2 = (ba^3)a^2 = b(a^3a^2) = ba$$

## ■ Example 13.2.2

Find all the subgroups of the dihedral group $D_4$.

First, there are the trivial subgroups, $\{e\}$ and $D_4$ itself. Then there is the subgroup $\{e, a^2\}$, which you can think of as the half-turn subgroup. The subgroup $\{e, a, a^2, a^3\}$ completes those that can be made using only $e$ and powers of $a$.

There are four other subgroups of order 2. They are $\{e, b\}$, $\{e, ba\}$, $\{e, ba^2\}$ and $\{e, ba^3\}$.

Finally, there are the subgroups $\{e, a^2, b, ba^2\}$ and $\{e, a^2, ba, ba^3\}$.

> *These are all the subgroups of $D_4$, but no proof is given at this stage. The subgroups of $D_n$ are the subject of Section 13.4.*

## 13.3 THE GENERAL DIHEDRAL GROUP $D_n$

Suppose that $A_1A_2 \ldots A_n$ is a regular $n$-sided polygon. Let $a$ be the symmetry operation of rotating $A_1A_2 \ldots A_n$ through $2\pi/n$ radians about its centre, and let $b$ be any reflection in an axis of symmetry. Fig. 13.6 shows the case when the axis of symmetry of $b$ passes through a vertex of the regular $n$-sided polygon.

The elements $e$, $a$, $a^2$, ... , $a^{n-1}$ are all distinct, and $b$ is not equal to any of them. Therefore the $2n$ elements $e$, $a$, $a^2$, ... , $a^{n-1}$, $b$, $ba$, $ba^2$, ... $ba^{n-1}$ are all distinct. (See question 5, Exercise 13.)

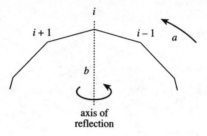

**Fig. 13.4**

Also the number of elements in $D_n$ is $2n$. For whether $n$ is even or odd, there are $n$ rotational symmetries. If $n$ is even there are $\frac{1}{2}n$ axes of reflection which do not pass through a vertex, and $\frac{1}{2}n$ axes which do, that is $n$ axes of reflection in all; when $n$ is odd, there are $n$ axes of reflectional symmetry which pass through a vertex and through a mid-point of a side. So, there are always a total of $2n$ symmetries in $D_n$.

As the $2n$ elements $e, a, a^2, ..., a^{n-1}, b, ba, ba^2, ..., ba^{n-1}$ are distinct and $D_n$ has no more than $2n$ elements, it follows that $D_n = \left\{ e, a, a^2, ..., a^{n-1}, b, ba, ba^2, ..., ba^{n-1} \right\}$. This generalises the particular case results of $D_3$, and $D_4$.

From Fig. 13.4 you can see that $aba = b$. You can also check that $aba = b$ for the case when the axis of symmetry passes through the mid-point of a side and does not pass through a vertex.

So for the dihedral group $D_n$, you now have $a^n = b^2 = e$ and $aba = b$. This enables you to prove, by induction on $i$, that $a^i ba^i = b$ for all $i$, which enables you to do calculations in $D_n$. For example, in $D_6$, to find $ba^4 \circ ba^5$, you write:

$$ba^4 \circ ba^5 = ba^4 ba^5 = b\left(a^4 ba^4\right)a = b(b)a = b^2 a = a$$

## 13.4 SUBGROUPS OF DIHEDRAL GROUPS

### ■ *Example 13.4.1*

Find the subgroups of $D_{12}$.

Figure 13.5 shows a regular dodecagon, with its two inscribed regular hexagons, its three inscribed squares and its four inscribed equilateral triangles.

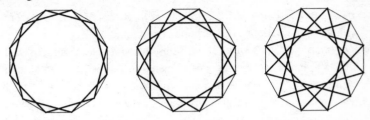

**Fig. 13.5**

For each of these inscribed regular polygons, its group of symmetries is a subgroup of $D_{12}$. Moreover, these subgroups are all different.

For consider two of the squares shown in Fig. 13.6, with vertices numbered 1, 4, 7, 10 and 2, 5, 8, 11.

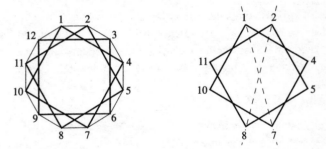

**Fig. 13.6**

If $L$ is the reflection in the axis through points 1 and 7, and $M$ is the reflection in the axis through points 2 and 8, $L$ is a symmetry of the square 1, 4, 7, 10 but not of square 2, 5, 8, 11, and vice-versa for $M$.

So inside $D_{12}$ there are two copies of $D_6$, three copies of $D_4$ and four copies of $D_3$. There are also six copies of $D_2$, one for each pair of orthogonal axes. Each of these $D_2$s is of the form $\{I, L, M, LM\}$ where $L, M$ are the reflections, and $LM$ is a rotation through 180°.

There are also 12 copies of $D_1$. Each of these is a subgroup of order 2 consisting of $I$ together with one of the 12 reflections.

In addition to these 28 dihedral subgroups of $D_{12}$, (27 plus $D_{12}$ itself), there are also subgroups consisting of rotations only.

There are six of these:

- the group consisting of the rotations of multiples of $\frac{1}{6}\pi$. There are twelve such rotations. This group is isomorphic to $\mathbf{Z}_{12}$
- the group consisting of all those rotations which are symmetries of one, and hence all, the inscribed hexagons, that is, the rotations of multiples of $\frac{1}{3}\pi$. This group is isomorphic to $\mathbf{Z}_6$
- the group consisting of all those rotations which are symmetries of one, and hence all, the inscribed squares, that is, the rotations of multiples of $\frac{1}{4}\pi$. This group is isomorphic to $\mathbf{Z}_4$
- the group consisting of all those rotations which are symmetries of one, and hence all, the inscribed equilateral triangles, that is, the rotations of multiples of $\frac{2}{3}\pi$. This group is isomorphic to $\mathbf{Z}_3$
- the group $\{I, P\}$, where $P$ is the rotation of $\pi$
- the identity subgroup $\{I\}$.

These subgroups completely describe all the subgroups of $D_{12}$, but a proof that there are no other subgroups of $D_{12}$ is not available until after Theorem 48 (in the next chapter) which states that, for a finite group, the order of a subgroup must divide the order of the group.

## ■ *Example 13.4.2*

The particular case which was discussed in Example 13.4.1 should enable you to follow the general case of subgroups of the dihedral group $D_n$. The results are given without proof.

$$D_n = \left\{e, a, a^2, \dots, a^{n-1}, b, ba, ba^2, \dots, ba^{n-1}\right\}$$

Again, there are cyclic subgroups and dihedral subgroups. Each of the subgroups of $\left\{e, a, a^2, \dots, a^{n-1}\right\}$ is a subgroup of $D_n$, and, by Theorem 24, all of these are cyclic. Looking ahead to Theorem 51, for every divisor of $n$ there is precisely one cyclic group of order $d$, namely, $\left\{e, a^s, a^{2s}, \dots, a^{(d-1)s}\right\}$, where $s = n/d$.

For each divisor $d$ of $n$, there are also $n/d$ dihedral subgroups of $D_n$, these being:

$$\left\{e, a^s, a^{2s}, \dots, a^{(d-1)s}\right\} \cup \left\{b, ba^s, ba^{2s}, \dots, ba^{(d-1)s}\right\}$$

$$\left\{e,a^s,a^{2s},\ldots,a^{(d-1)s}\right\}\cup\left\{ba,ba^{s+1},ba^{2s+1},\ldots,ba^{(d-1)s+1}\right\}$$

$$\vdots$$

$$\left\{e,a^s,a^{2s},\ldots,a^{(d-1)s}\right\}\cup\left\{ba^{s-1},ba^{2s-1},ba^{3s-1},\ldots,ba^{ds-1}\right\}$$

So, for each divisor $d$ of $n$, there is one cyclic subgroup and $n/d$ dihedral subgroups of $D_n$.

> *You need a little care in interpreting this statement when $n$ is even. For in this case, there are $n$ dihedral groups of order 2, and these are all cyclic!*

Are there any other subgroups of $D_n$?

Suppose that $A$ is a subgroup of $D_n$. Write $C_n=\left\{e,a,a^2,\ldots,a^{n-1}\right\}$. If $A\subseteq C_n$, then, by Theorem 24, $A$ is cyclic.

If $A\not\subseteq C_n$, then $ba^m\in A$ for some $m$. Let $B=A\cap C_n$. Then $B$ is a subgroup of $C_n$, so $B$ is cyclic; therefore $B=\left\{e,a^s,a^{2s},\ldots,a^{(d-1)s}\right\}$ for some $s$, $d$ with $sd=n$, looking forward again to Theorem 51.

Then you can show, see Exercise 13, question 7, that:

$$A=\left\{e,a^s,a^{2s},\ldots,a^{(d-1)s}\right\}\cup\left\{ba^m,ba^{m+s},ba^{m+2s},\ldots,ba^{m+(d-1)s}\right\}$$

This is one of the groups listed above, which shows that $A$ is one of the $n/d$ dihedral subgroups of $D_n$.

## WHAT YOU SHOULD KNOW

- The properties of $D_n$ and how to carry out calculations in it.

## EXERCISE 13

**1** Verify that $aba=b$ when the axis of reflection of a regular polygon passes through the mid-point of a side. Use the relation $aba=b$ to show that $(ba)^2=e$, and prove more generally that $\left(ba^i\right)^2=e$ for $i=0,1,\ldots,n-1$.

**2** Carry out the following calculations in $D_n$, giving your answers in the form $b^i a^j$, where $i\in\{0,1\}$ and $j\in\{0,1,\ldots,n-1\}$.

(1) $a(ba)$,  (2) $a^{-1}$,  (3) $(ba)^{-1}$,

(4) $bab^{-1}a^{-1}$,  (5) $(ba)(ba^2)$.

**3**  Prove that, in $D_n$, $a^i b = b a^{n-i}$, for $i \in \{1, 2, 3, \ldots, n-1\}$.

**4**  How many subgroups of order 2 are there in $D_n$?

**5**  Show that in $D_n$, if $ba^i = ba^j$ where $i, j \in \{0, 1, 2, \ldots, n-1\}$, then $i = j$, and that $ba^i \neq a^j$ for any $i$ or $j$.

**6**  Figure 13.7 shows a regular tetrahedron. The group $G$ of rotational symmetries of the regular tetrahedron is called the tetrahedral group.

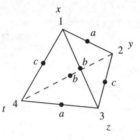

**Fig. 13.7** *A regular tetrahedron*

In Fig. 13.7, $x$, $y$, $z$ and $t$ refer to the fixed axes from the vertices of the tetrahedron to the centres of the opposite faces. $X$, $Y$, $Z$ and $T$ are rotations of $\frac{2}{3}\pi$ clockwise about $x$, $y$, $z$ and $t$ respectively. The two points marked $a$ are opposite ends of a fixed axis in space passing through the mid-points of opposite edges; $A$ is a half-turn about $a$; similarly $B$ and $C$ are half-turns about corresponding axes indicated by $b$ and $c$. Write down all the different rotational symmetries of the regular tetrahedron, and find their orders. Show that there is no subgroup of order 6. Write down the image of each element of $G$ under the function $f : G \to S_4$ defined in a similar way to that in Section 13.1, and hence show that $G$ is isomorphic to $A_4$.

**7**  Let $A$ be a subgroup of $D_n$ such that $ba^m \in A$ for some $m$, and $A \cap C_n = \left\{ e, a^s, a^{2s}, \ldots, a^{((d-1)s)} \right\}$, where $sd = n$. Prove the result mentioned at the end of Example 13.4.2, namely that:

$$A = \left\{ e, a^s, a^{2s}, \ldots, a^{(d-1)s} \right\} \cup \left\{ ba^m, ba^{m+s}, ba^{m+2s}, \ldots, ba^{m+(d-1)s} \right\}$$

# 14

## *Cosets*

## 14.1 INTRODUCTION

You have almost certainly observed that the number of elements of a subgroup of a finite group is a factor of the number of elements in the group.

This is an important result, called Lagrange's theorem.

This chapter is devoted to proving Lagrange's theorem and, on the way, sets up some important machinery to be used later in the book.

The important idea is that of a coset.

## 14.2 COSETS

**Definition**: Let $H$ be a subgroup of a group $G$, and let $x \in G$. Then the set of elements $xH$ defined by $xH = \{xh : h \in H\}$ is called a **left coset** of $H$ (in $G$). The set $Hx$ defined by $Hx = \{hx : h \in H\}$ is called a **right coset** of $H$ (in $G$). $xH$ is the left coset of $H$ containing (or generated by) $x$.

## ■ *Example 14.2.1*

As an example, consider the group $D_3$ and the subgroup $H = \left\{e, a, a^2\right\}$.

|       | $e$    | $a$    | $a^2$  | $b$    | $ba$   | $ba^2$ |
|-------|--------|--------|--------|--------|--------|--------|
| $e$   | $e$    | $a$    | $a^2$  | $b$    | $ba$   | $ba^2$ |
| $a$   | $a$    | $a^2$  | $e$    | $ba^2$ | $b$    | $ba$   |
| $a^2$ | $a^2$  | $e$    | $a$    | $ba$   | $ba^2$ | $b$    |
| $b$   | $b$    | $ba$   | $ba^2$ | $e$    | $a$    | $a^2$  |
| $ba$  | $ba$   | $ba^2$ | $ba$   | $a^2$  | $e$    | $a$    |
| $ba^2$| $ba^2$ | $b$    | $ba$   | $a$    | $a^2$  | $e$    |

**Fig 14.1**

Then the six possible left cosets are:

$$eH = \left\{ee, ea, ea^2\right\} = \left\{e, a, a^2\right\}$$
$$aH = \left\{ae, aa, aa^2\right\} = \left\{a, a^2, e\right\}$$
$$a^2 H = \left\{a^2 e, a^2 a, a^2 a^2\right\} = \left\{a^2, e, a\right\}$$
$$bH = \left\{be, ba, ba^2\right\} = \left\{b, ba, ba^2\right\}$$
$$baH = \left\{bae, baa, baa^2\right\} = \left\{ba, ba^2, b\right\}$$
$$ba^2 H = \left\{ba^2 e, ba^2 a, ba^2 a^2\right\} = \left\{ba^2, b, ba\right\}.$$

> *Notice that the first three cosets are equal to each other, as are the second three.*

## ■ *Example 14.2.2*

Write out the left cosets of the subgroup $3\mathbf{Z}$ of $\mathbf{Z}$ under addition.

> *In this example additive notation is used for cosets. Notice also that, for abelian groups, left cosets are the same as right cosets.*

The coset corresponding to 0 is the set $0+3\mathbf{Z}$, which is the set formed by adding every element of $3\mathbf{Z}$, that is the set of multiples of 3, in turn, to 0. Hence $0+3\mathbf{Z}=3\mathbf{Z}$. See Example 4.4.3. Other cosets are $1+3\mathbf{Z}=\{...,-5,-2,1,4,7,...\}$ and $2+3\mathbf{Z}=\{...,-4,-1,2,5,8,...\}$. Now consider the coset of $r$, that is $r+3\mathbf{Z}$. If $r\equiv 0\,(\mathrm{mod}\,3)$, then $r+3\mathbf{Z}=3\mathbf{Z}$. If $r\equiv 1\,(\mathrm{mod}\,3)$, then $r+3\mathbf{Z}=1+3\mathbf{Z}$. And, if $r\equiv 2\,(\mathrm{mod}\,3)$, then $r+3\mathbf{Z}=2+3\mathbf{Z}$. There are thus just three cosets of $3\mathbf{Z}$ of $\mathbf{Z}$.

*Look at the cosets in these two examples carefully, and write down any observations that you have about them. You are also advised to write out in a similar way the cosets of the subgroup $H=\{e,x\}$ in $D_3$ and see whether your observations hold for these cosets before reading on.*

*Theorems 46 and 47 prove generally results which you may have suspected from these examples.*

**Theorem 46** Let $H$ be a subgroup of a group $G$ and let $x,y\in H$. Then:

(1) $xH=yH$ if, and only if, $x^{-1}y\in H$

(2) $x\in$ coset $A$ if, and only if, $A=xH$

(3) $xH=yH$ if, and only if, $x$ and $y$ are in the same coset.

**Proof**: (1) *If.* Suppose that $x^{-1}y\in H$, so $x^{-1}y=h$ for some $h$ in $H$. Then $y=xh$ and $x=yh^{-1}$.

*To prove this part of the theorem you need to show that if $x^{-1}y\in H$ then $xH=yH$. To do this you need to show that $xH\subseteq yH$ and $yH\subseteq xH$. So start by taking an element $a\in xH$ and show that it belongs to $yH$; then take an element $a\in yH$ and show that it belongs to $xH$.*

Suppose that $a\in xH$. Then $a=xh_1$ for some $h_1\in H$. Therefore, since $x=yh^{-1}$, you can write $a$ in the form $a=xh_1=yh^{-1}h_1$. Since $H$ is a subgroup, $h^{-1}h_1\in H$ from Theorem 21. Therefore $a=y\big(h^{-1}h_1\big)\in yH$, so $xH\subseteq yH$.

Suppose now that $a \in yH$. Then $a = yh_1$ for some $h_1 \in H$. Therefore, since $y = xh$, you can write $a$ in the form $a = yh_1 = xhh_1$. Since $H$ is a subgroup, $hh_1 \in H$ from Theorem 21. Therefore $a = x(hh_1) \in xH$, so $yH \subseteq xH$.

Therefore, as $xH \subseteq yH$ and $yH \subseteq xH$, $xH = yH$.

*Only if.* Suppose that $xH = yH$. As $y = ye$ and $e \in H$ because $H$ is a subgroup, $y$ is in the coset $yH$. Therefore, as $xH = yH$, $y \in xH$, so $y = xh$ for some element $h$ of $H$. Therefore $x^{-1}y = h$, so $x^{-1}y \in H$.

(2) *If.* Suppose the coset $A = xH$. Then, as $e \in H$, $xe \in xH$ so $x \in xH$ and hence $x \in A$.

*Only if.* Suppose that $x \in A$, where $A$ is a coset. Suppose that $A$ is the coset corresponding to $y$, that is $A = yH$. If $x \in yH$, then $x = yh$ for some $h \in H$. Therefore $y^{-1}x = h$, so $y^{-1}x \in H$. But, from part (1), with the roles of $x$ and $y$ interchanged, $xH = yH$, so $A = xH$. So, if $x$ belongs to a coset, that coset is $xH$.

(3) *If.* Suppose that $x$ and $y$ are in the same coset $A$. By part (2), $A = xH$ and $A = yH$, so $xH = yH$.

*Only if.* Suppose that $xH = yH$. Then, from part (1), $x^{-1}y \in H$, so $x^{-1}y = h$ for some $h \in H$. Therefore $y = xh$, so $y \in xH$. But $x \in xH$, from part (2), so $x$ and $y$ are in the same coset. ∎

It follows immediately by putting $x = e$ in the statement of Theorem 46, part (1), that $yH = H$ if, and only if, $y \in H$.

**Theorem 47**: Let $H$ be a subgroup of a group $G$. Then $G$ is a disjoint union of left cosets of $H$ in $G$.

**Proof**: For each element $x$ of the group $G$, $x \in xH$, because $x = xe$ and $e \in H$. Therefore $G$ is a union of left cosets of $H$ and it remains to show that distinct cosets are disjoint.

If $xH \cap yH$ is non-empty, then $xh_1 = yh_2$ for some $h_1, h_2 \in H$. Then $x^{-1}y = h_1h_2^{-1} \in H$, so that $xH = yH$ by Theorem 46. Therefore $xH$ and $yH$ are either the same or disjoint. ∎

## 14.3 LAGRANGE'S THEOREM

It is now time to prove Lagrange's theorem. The proof follows quickly from Theorem 47.

**Theorem 48: Lagrange's theorem**: Let $H$ be a subgroup of a finite group $G$. Then the order of $H$ divides the order of $G$.

**Proof**: Suppose that the order of $G$ is $n$, and the order of $H$ is $m$. Then each left coset of $H$ has exactly $m$ elements, for if $H = \{h_1, h_2, \ldots, h_m\}$ then $xH = \{xh_1, xh_2, \ldots, xh_m\}$. (Notice that no two elements of the form $xh_i$ and $xh_j$ $(i \neq j)$ are equal, because if $xh_i = xh_j$ then $h_i = h_j$, by Theorem 15.)

So, by Theorem 47, $G$ is a disjoint union of subsets each containing $m$ elements. Therefore $m$ divides $n$. ∎

You may very well wonder whether the converse of Lagrange's theorem is true, that is, if m is a factor of the order of a group, that there will be a subgroup of order n. The answer is no; the alternating group $A_4$ is a group of order 12 which has no subgroup of order 6. (Recall question 6 in Chapter 13.) However, the converse is true for finite cyclic groups, as you will see from Theorem 52 in the next section.

## 14.4 DEDUCTIONS FROM LAGRANGE'S THEOREM

There are several corollaries of Lagrange's theorem.

**Theorem 49**: In a finite group $G$, the order of an element of $G$ divides the order of $G$.

**Proof**: The order $m$ of $g \in G$ is the order of the subgroup generated by $g$. Hence, by Lagrange's theorem, $m$ divides the order of $G$. ∎

From this result you can deduce a number of results simply from knowing the order of a group. For example, if a group has five elements, any subgroup can only have one element, and must therefore be the subgroup consisting of the identity alone, or have five elements, in which case it is the whole group. Moreover, the only element of order one is the identity element, and that is unique, so the order of every other element of a group with five elements is five. Therefore a group with five elements is cyclic.

This leads to the following theorem.

**Theorem 50**: Every group of prime order is cyclic.

**Proof**: Suppose that the order of a group is a prime $p$. Consider an element $g$ which is not the identity. Since the order of $g$ divides $p$ and is not 1, it must be $p$ itself. Therefore $g$ is a generator of the group, so the group is cyclic. ∎

> *Together with Theorem 36, this result says, in effect, that given a prime $p$, there is only one group of order $p$, namely the cyclic group of order $p$.*

Here are two other results which follow from Lagrange's theorem.

**Theorem 51**: Let $G$ be a finite group of order $n$. Then $x^n = e$ for all $x \in G$.

**Proof**: From Theorem 49 the order of each element of $G$ divides $n$. Suppose that the order of an element $x \in G$ is $m$. Then $m$ divides $n$, so $n = km$ for some positive integer $k$. It follows that $x^n = x^{km} = \left(x^m\right)^k = e^k = e$. ∎

**Theorem 52**: Let $G$ be a finite cyclic group of order $n$. Then for every divisor $d$ of $n$ there is exactly one subgroup of order $d$.

**Proof**: Since $G$ has order $n$, you can write $G = \left\{e, a, a^2, \ldots, a^{n-1}\right\}$. Let $d$ be a divisor of $n$. Then $H = \left\{e, a^{n/d}, a^{2n/d}, \ldots, a^{(d-1)n/d}\right\}$ is a subgroup of $G$ of order $d$.

> *This shows that there is a subgroup of $G$ of order $d$. Now you have to show that there are no others.*

Let $K$ be a subgroup of $G$ of order $d$. If $b \in K$ then $b = a^s$, for some $s$ and $b^d = e$, by Theorem 51. Therefore $\left(a^s\right)^d = e$, so $a^{sd} = e$. But then, from Theorem 17, part (2), $n$ divides $sd$, so $s = m(n/d)$ for some

$m \in \mathbf{Z}$. Therefore $b = a^s = \left(a^{n/d}\right)^m$, so $b \in H$. Therefore $K \subseteq H$. But then $H$ and $K$ both have the same number of elements, so $H = K$. ∎

## 14.5 TWO NUMBER THEORY APPLICATIONS

Theorem 52 has two applications in number theory.

*If you wish, you could omit these two theorems, as future work does not depend on them. However, you may wish to note the results.*

**Theorem 53: Fermat's little theorem**: Let $p$ be a prime number. Then $a^p \equiv a \pmod{p}$ for all $a \in \mathbf{Z}$.

**Proof**: Consider the group $\left(\mathbf{Z}_p{}^*, \times\right)$. This group has $p-1$ elements. For $[a] \in \mathbf{Z}_p{}^*$, by Theorem 51, $\left[a^{p-1}\right] = [a]^{p-1} = [1]$, the identity in $\left(\mathbf{Z}_p{}^*, \times\right)$. Therefore $a^{p-1} \equiv 1 \pmod{p}$ for all integers $a$ not divisible by $p$. Therefore $a^p \equiv a \pmod{p}$ for all $a \in \mathbf{Z}$. (Note that it is easy to see that $a^p \equiv a \pmod{p}$ when $a$ is divisible by $p$.) ∎

**Theorem 54**: The group $\left(\mathbf{Z}_p{}^*, \times\right)$ is cyclic.

**Proof**: First, if $m$ is the largest of the orders of the elements of $\mathbf{Z}_p{}^*$, then $a^m = 1$ for all $a \in \mathbf{Z}_p{}^*$. For if $n$ is the order of the element $a$, by Theorem 20, there is an element which has as its order the least common multiple of $m$ and $n$. Suppose that $l$ is the least common multiple of $m$ and $n$. Then $m$ divides $l$. But $l \le m$, because one of the elements has order $l$, and $m$ is the biggest of the orders of the elements. Therefore $l = m$. Therefore $n$ divides $m$, and $a^m = 1$.

Let $m$ be the largest of the orders of the elements of $\mathbf{Z}_p{}^*$. Then, from the previous paragraph, each of the $p-1$ elements of $\mathbf{Z}_p{}^*$ satisfies the polynomial equation $x^m = 1$. But, by Theorem 10, a polynomial of degree $m$ with coefficients in $\mathbf{Z}_p$ can have no more than $m$ roots. Therefore $p-1 \le m$.

On the other hand, from Theorem 49, the first deduction from Lagrange's theorem, the order of an element divides the order of the group, so $m$ divides $p-1$. Therefore $m = p-1$.

Thus there is an element of order $p-1$, so $\left(\mathbf{Z}_p{}^*, \times\right)$ is cyclic. ∎

## ■ *Example 14.5.1*

The group $(\mathbf{Z}_7{}^*, \times)$ consists of the elements $\{1, 2, 3, 4, 5, 6\}$ and, according to Theorem 54, it is cyclic and therefore has a generator. However, it may not be obvious which element is a generator. In this case, $3^2 = 2$, $3^3 = 6$, $3^4 = 4$, $3^5 = 5$ and $3^6 = 1$. The element 5 is also a generator.

On the other hand, in $(\mathbf{Z}_{101}{}^*, \times)$ which has 100 elements, it may take some time to find a generator. Theorem 54, however, guarantees that it does have one.

## 14.6 MORE EXAMPLES OF COSETS

## ■ *Example 14.6.1*

Let $G = \{1, \theta, \theta^2, \theta^3, \theta^4, \theta^5\}$ be the group consisting of the six complex sixth roots of unity, where $\theta = e^{\pi i/3}$, and let $H = \{1, \theta^2, \theta^4\}$. Write out the left cosets of $H$.

The left cosets of $H$ are $H = \{1, \theta^2, \theta^4\}$ and $\theta H = \{\theta, \theta^3, \theta^5\}$. These cosets partition the group, so any other cosets must be identical to one or other of these. In fact, $H = \theta^2 H = \theta^4 H = \{1, \theta^2, \theta^4\}$ and $\theta H = \theta^3 H = \theta^5 H = \{\theta, \theta^3, \theta^5\}$.

## ■ *Example 14.6.2*

Consider the group $(\mathbf{C}^*, \times)$ and the subgroup $T = \{z \in \mathbf{C}^* : |z| = 1\}$.

For $a \in \mathbf{C}^*$, the coset $aT = \{az \in \mathbf{C}^* : |z| = 1\}$, which is the same as $\{z \in \mathbf{C}^* : |z| = |a|\}$.

*You can also think of this geometrically. In the complex plane, the subgroup T is the set of points on the unit circle, with the operation of multiplication in the complex plane. To find the coset which includes another point a in the plane, multiply a by each point in the unit circle, and you get the circle of radius |a|. The cosets are therefore the circles of radius |a| in the complex plane. Fig. 14.1 shows the complex plane, the unit circle T, and some of the circles which are cosets. Note that the cosets are disjoint, and their union is* **C***.*

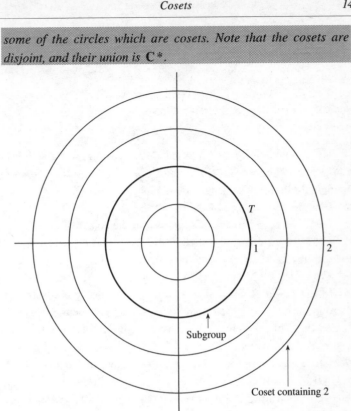

**Fig. 14.1** *The unit circle subgroup of* $(\mathbf{C}^*, \times)$*, and its cosets*

## WHAT YOU SHOULD KNOW

- The meaning of left and right cosets.

- Given a subgroup $H$ of a group $G$, every element of $G$ is in some left coset of $H$.

- That for any subgroup $H$ of a group $G$, $G$ is a union of disjoint cosets of $H$.

- That $x^{-1}y \in H$ is a necessary and sufficient condition for the cosets $xH$ and $yH$ to be identical.

■ In a finite group all the cosets of a given subgroup have the same number of elements.

■ Lagrange's theorem.

■ Every group of prime order is cyclic.

## EXERCISE 14

**1** Use the table in Fig. 13.3 to write out the left cosets and the right cosets of the subgroups $\{e, a^2\}$ and $\{e, b\}$ in $D_4$. Write down one important difference between the two cases.

**2** Find the left cosets of $\mathbf{Z}_2 \times \{0\}$ in the group $\mathbf{Z}_2 \times \mathbf{Z}_3$.

**3** Let $G = (\mathbf{Z}, +)$ and $H = 4\mathbf{Z}$. Write out the cosets of $H$.

**4** Let $H$ be a subgroup of a group $G$. Prove that there is a bijection between the left cosets of $H$ and the right cosets of $H$.

**5** $(\mathbf{Z}, +)$ is a subgroup of $(\mathbf{R}, +)$. What are the left cosets of $\mathbf{Z}$?

**6** In the group $\mathbf{R} \times \mathbf{R}$ under addition, show that the cosets of the subgroup formed by multiples of a fixed non-zero element $(a, b)$, which you can think of as vectors lying on a line through the origin, are the lines in $\mathbf{R} \times \mathbf{R}$ parallel to $(a, b)$.

**7** Let $H$ be a subgroup of a group $G$. Prove that if $x \in yH$ then $xH = yH$.

**8** Verify that $(\mathbf{Z}_{13}{}^*, \times)$ is a cyclic group by finding a generator.

**9** Mark each of the following statements true or false.

(1) Every subgroup of every group is a coset of that group.

(2) You cannot have an infinite number of cosets in a group.

(3) In an infinite group, you cannot have both an infinite subgroup and an infinite number of cosets.

(4) If two cosets have a common element, they are identical.

**10** Let $H$ and $K$ be finite subgroups of a group $G$, and let the orders of $H$ and $K$ be relatively prime. Prove that $H \cap K = \{e\}$.

**11** Show that if $H$ and $K$ are subgroups of a group, and have orders 56 and 63 respectively, then the subgroup $H \cap K$ must be cyclic.

# 15

# Groups of orders up to 8

## 15.1 INTRODUCTION

This chapter is different from others in this book. Its purpose is to find all the groups with orders up to 8. The main tool is Theorem 48, Lagrange's theorem, which states that the order of an element must divide the order of the group.

Although all cyclic groups of order $n$ are isomorphic to $\mathbf{Z}_n$, this group is, nevertheless, a specific group $\{[0], [1], \ldots, [n-1]\}$ of residue classes of integers under addition. There is sometimes an advantage in thinking abstractly in terms of $C_n = \{e, a, a^2, \ldots, a^{n-1}\}$ which can legitimately 'stand for' any cyclic group of order $n$. In this chapter, the notation $C_n$ is used for the cyclic group of order $n$.

## 15.2 GROUPS OF PRIME ORDER

From Theorem 50, every group of prime order is cyclic. This shows that, up to isomorphism, the only groups of orders 2, 3, 5 and 7 are the cyclic groups $C_2$, $C_3$, $C_5$ and $C_7$.

## 15.3 GROUPS OF ORDER FOUR

Let $G$ be a group of order 4. Think of the orders of the elements of $G$. Each of these divides the order of $G$, so, apart from the identity element, these orders can be only 2 or 4.

First suppose that there is an element of order 4. In this case $G$ must be the cyclic group $C_4$.

Suppose, on the other hand, that every non-identity element is of order 2, and let $a$ and $b$ be two distinct (non-identity) elements. Consider the element $ba$. It is not the identity, for if it were then $b = a^{-1} = a$; it is not $a$ or $b$, because, if so, $b = e$ or $a = e$. The element $ba$ is therefore not equal to any of the other elements, and $G = \{e, a, b, ba\} \cong C_2 \times C_2$.

You can also see this by applying Theorem 38.

The group table for $C_2 \times C_2$ is shown in Fig. 15.1. This group is the four-group, $V$, which you met in question 4 of Exercise 5.

|      | $e$  | $a$  | $b$  | $ba$ |
|------|------|------|------|------|
| $e$  | $e$  | $a$  | $b$  | $ba$ |
| $a$  | $a$  | $e$  | $ba$ | $b$  |
| $b$  | $b$  | $ba$ | $e$  | $a$  |
| $ba$ | $ba$ | $b$  | $a$  | $e$  |

**Fig. 15.1** *The group* $V \cong C_2 \times C_2$

## 15.4 GROUPS OF ORDER SIX

Let $G$ be a group of order 6. Then the orders of the elements must divide six, so, apart from the identity element, these orders can be only 2, 3 or 6.

It is not possible for all the elements of $G$ to have order 2. This follows from Theorem 38, as 6 is not a power of 2.

First suppose that $G$ has an element of order 6. In this case $G$ must be the cyclic group $C_6$.

Now suppose that $G$ has an element of order 3, but no element of order 6. Call this element $a$.

If $b \notin \{e, a, a^2\}$, then $G = \{e, a, a^2, b, ba, ba^2\}$, as these are six distinct elements of $G$. Notice that $b^2$ cannot be equal to any of the last four members of $G$, because that would mean that $b$ is a power of $a$. Nor can $b^2$ be $a$ or $a^2$, because that would imply that $b$ is of order 6. This leaves only $b^2 = e$.

The strategy now is to look at the possible outcomes of the product $ab$. Clearly, $ab \neq e, a, a^2$ or $b$. So there are now two sub-cases to consider: $ab = ba$ and $ab = ba^2$.

**Case 1**: $ab = ba$

Considering powers of $ab$ gives the following results.

$$(ab)^2 = abab = aabb = a^2$$
$$(ab)^3 = (ab)^2 ab = a^2 ab = b$$
$$(ab)^4 = ab(ab)^3 = abb = a$$
$$(ab)^5 = (ab)^4 ab = aab = a^2 b$$
$$(ab)^6 = (ab)^5 ab = a^2 bab = a^2 abb = a^3 b^2 = e$$

This shows that the order of $ab$ is 6. This contradicts the hypothesis that there is no element of order 6.

**Case 2**: $ab = ba^2$

In this case, since $ab = ba^2$, $aba = ba^2 a = b$. You now have the relations $a^3 = b^2 = e$ and $aba = b$. The group $G$ is therefore isomorphic to $D_3$. Its group table is shown in Fig. 5.3.

## 15.5 GROUPS OF ORDER EIGHT

Let $G$ be a group of order 8. If $G$ has an element of order 8, then $G$ is the cyclic group $C_8$.

If $G$ is not cyclic, then the order of every element, except the identity, must be a proper divisor of 8, and so be 2 or 4.

Suppose that every non-identity element is of order 2. Then Theorem 38 applies, so $G$ is isomorphic to $C_2 \times C_2 \times C_2$.

If $\{a,b,c\}$ is a minimal generating set for $G$, then, as in Theorem 37, $G = \{e,a,b,c,bc,ca,ab,abc\}$. The group table for $G$ is shown in Fig. 15.2.

|     | $e$   | $a$  | $b$  | $c$  | $bc$ | $ca$ | $ab$ | $abc$ |
|-----|-------|------|------|------|------|------|------|-------|
| $e$   | $e$   | $a$  | $b$  | $c$  | $bc$ | $ca$ | $ab$ | $abc$ |
| $a$   | $a$   | $e$  | $ab$ | $ca$ | $abc$| $c$  | $b$  | $bc$  |
| $b$   | $b$   | $ab$ | $e$  | $bc$ | $c$  | $abc$| $a$  | $ca$  |
| $c$   | $c$   | $ca$ | $bc$ | $e$  | $b$  | $a$  | $abc$| $ab$  |
| $bc$  | $bc$  | $abc$| $c$  | $b$  | $e$  | $ab$ | $ca$ | $a$   |
| $ca$  | $ca$  | $c$  | $abc$| $a$  | $ab$ | $e$  | $bc$ | $b$   |
| $ab$  | $ab$  | $b$  | $a$  | $abc$| $ca$ | $bc$ | $e$  | $c$   |
| $abc$ | $abc$ | $bc$ | $ca$ | $ab$ | $a$  | $b$  | $c$  | $e$   |

**Fig. 15. 2** *The group* $C_2 \times C_2 \times C_2$

The remaining cases concern groups in which every element except the identity is of order 2 or 4, and at least one element, $a$ say, has order 4.

If $b \notin \{e,a,a^2,a^3\}$, then $G = \{e,a,a^2,a^3,b,ba,ba^2,ba^3\}$, as these are eight distinct elements of $G$. Notice that $b^2$ cannot be equal to any of the last four members of $G$ in this list, because that would mean that $b$ is a power of $a$. Nor can $b^2$ be $a$ or $a^3$, because that would imply that $b$ is of order 8. So just two cases remain: $b^2 = e$ and $b^2 = a^2$.

**Case 1**: $b^2 = e$

The strategy now is to look at the possible outcomes of the product $ab$. Clearly, $ab \neq e, a, a^2, a^3$ or $b$. Also $ab \neq ba^2$, as $a = ba^2b^{-1}$

gives $a^2 = \left(ba^2b^{-1}\right)^2 = ba^4b^{-1} = bb^{-1} = e$. So there are now two subcases to consider: $ab = ba$ and $ab = ba^3$.

**Case 1.1**: $ab = ba$

In this case $G$ is abelian, and is isomorphic to $C_4 \times C_2$. Fig. 15.3 shows its group table. It is generated by $a$ and $b$ with the relations $a^4 = b^2 = e$ and $ba = ab$.

> You could use Theorem 39 to show that there are three abelian groups of order 8; these are $C_2 \times C_2 \times C_2$, $C_4 \times C_2$ and $C_8$.

|        | $e$    | $a$    | $a^2$  | $a^3$  | $b$    | $ba$   | $ba^2$ | $ba^3$ |
|--------|--------|--------|--------|--------|--------|--------|--------|--------|
| $e$    | $e$    | $a$    | $a^2$  | $a^3$  | $b$    | $ba$   | $ba^2$ | $ba^3$ |
| $a$    | $a$    | $a^2$  | $a^3$  | $e$    | $ba$   | $ba^2$ | $ba^3$ | $b$    |
| $a^2$  | $a^2$  | $a^3$  | $e$    | $a$    | $ba^2$ | $ba^3$ | $b$    | $ba$   |
| $a^3$  | $a^3$  | $e$    | $a$    | $a^2$  | $ba^3$ | $b$    | $ba$   | $ba^2$ |
| $b$    | $b$    | $ba$   | $ba^2$ | $ba^3$ | $e$    | $a$    | $a^2$  | $a^3$  |
| $ba$   | $ba$   | $ba^2$ | $ba^3$ | $b$    | $a$    | $a^2$  | $a^3$  | $e$    |
| $ba^2$ | $ba^2$ | $ba^3$ | $b$    | $ba$   | $a^2$  | $a^3$  | $e$    | $a$    |
| $ba^3$ | $ba^3$ | $b$    | $ba$   | $ba^2$ | $a^3$  | $e$    | $a$    | $a^2$  |

**Fig. 15.3** *The group* $C_4 \times C_2$

**Case 1.2**: $ab = ba^3$

In this case, you now have the relations $a^4 = b^2 = e$ and $ab = ba^3$, which is equivalent to $aba = b$, as $a^3 = a^{-1}$. The group $G$ is therefore isomorphic to $D_4$. Its group table is shown in Fig. 13.3.

**Case 2**: $b^2 = a^2$

In this case, both $a$ and $b$ have order 4. As in case 1, it is clear that $ab \neq e, a, a^2, a^3$ or $b$. In addition, if $ab = ba^2$ then $ab = bb^2$ and $ab^2 = b^4 = e$ from which it would follow that $a = b^{-2} = b^2$, a contradiction. So there must again be two sub-cases, $ab = ba$ and $ab = ba^3$.

**Case 2.1**: $ab = ba$

In this case, $G$ is abelian. Put $c = ab^{-1}$. Then $c$ is of order 2, because $\left(ab^3\right)^2 = a^2 b^6 = a^8 = e$. In this case $G = \left\{e, a, a^2, a^3, c, ca, ca^2, ca^3\right\}$ which you can see is $C_4 \times C_2$ again.

**Case 2.2**: $ab = ba^3$

In this case, $G = \left\{e, a, a^2, a^3, b, ba, ba^2, ba^3\right\}$ where $a^4 = e$, $a^2 = b^2$ and $ab = ba^3$.

There is a group of matrices which has the properties above. If $A = \begin{pmatrix} i & 0 \\ 0 & -i \end{pmatrix}$ and $B = \begin{pmatrix} 0 & -1 \\ 1 & 0 \end{pmatrix}$, you can verify, using Theorem 21, that the set $\left\{I, A, A^2, A^3, B, BA, BA^2, BA^3\right\}$ is a subgroup of the group of invertible $2 \times 2$ matrices with complex entries under multiplication. Note also that $A^4 = I$, $A^2 = B^2$ and $AB = BA^3$.

It follows that the set $G = \left\{e, a, a^2, a^3, b, ba, ba^2, ba^3\right\}$ where $a^4 = e$, $a^2 = b^2$ and $ab = ba^3$ forms a group. This group is called the quaternion group $\mathbf{Q}_4$. Its group table is shown in Fig. 15.4.

|       | $e$    | $a$    | $a^2$  | $a^3$  | $b$    | $ba$   | $ba^2$ | $ba^3$ |
|-------|--------|--------|--------|--------|--------|--------|--------|--------|
| $e$    | $e$    | $a$    | $a^2$  | $a^3$  | $b$    | $ba$   | $ba^2$ | $ba^3$ |
| $a$    | $a$    | $a^2$  | $a^3$  | $e$    | $ba^3$ | $b$    | $ba$   | $ba^2$ |
| $a^2$  | $a^2$  | $a^3$  | $e$    | $a$    | $ba^2$ | $ba^3$ | $b$    | $ba$   |
| $a^3$  | $a^3$  | $e$    | $a$    | $a^2$  | $ba$   | $ba^2$ | $ba^3$ | $b$    |
| $b$    | $b$    | $ba$   | $ba^2$ | $ba^3$ | $a^2$  | $a^3$  | $e$    | $a$    |
| $ba$   | $ba$   | $ba^2$ | $ba^3$ | $b$    | $a$    | $a^2$  | $a^3$  | $e$    |
| $ba^2$ | $ba^2$ | $ba^3$ | $b$    | $ba$   | $e$    | $a$    | $a^2$  | $a^3$  |
| $ba^3$ | $ba^3$ | $b$    | $ba$   | $ba^2$ | $a^3$  | $e$    | $a$    | $a^2$  |

**Fig. 15.4** *The group* $\mathbf{Q}_4$

## 15.6 SUMMARY

### Groups of prime order $p$

The only groups are isomorphic to $C_p$.

### Groups of order 4

The only groups are isomorphic to:

- $C_4$, with generator $a$ such that $a^4 = e$

- $C_2 \times C_2$, with generators $a$ and $b$ such that $a^2 = b^2 = e$.

### Groups of order 6

The only groups are isomorphic to:

- $C_6$, with generator $a$ such that $a^6 = e$

- $D_3$, with generators $a$ and $b$ such that $a^3 = b^2 = e$ and $aba = b$.

### Groups of order 8

The only groups are isomorphic to:

- $C_8$, with generator $a$ such that $a^8 = e$

- $C_2 \times C_2 \times C_2$, with generators $a$, $b$ and $c$ such that $a^2 = b^2 = c^2 = e$

- $C_4 \times C_2$, with generators $a$ and $b$ such that $a^4 = b^2 = e$ and $ab = ba$

- $D_4$, with generators $a$ and $b$ such that $a^4 = b^2 = e$ and $aba = b$

- $Q_4$, with generators $a$ and $b$ such that $a^4 = e$, $a^2 = b^2$ and $ab = ba^3$.

## EXERCISE 15

**1** Use the methods of this chapter to analyse the groups which have

(1) order 11

(2) order 9

(3) order 10.

# 16

# *Equivalence relations*

## 16.1 INTRODUCTION

An equivalence relation is a mathematical way to describe 'sameness'.

Consider, for example, the set of lower and upper case letters of the alphabet, $\{a, b, c, \ldots A, B, C \ldots\}$. You may wish, for some purposes, to think of $e$ and $E$ as the 'same' as each other; for other purposes, you may wish to think of the lower case letters as the 'same' as each other and for yet other purposes, you may wish to think of vowels as the 'same' as each other. When two things are the 'same' in some way, in that they share some particular property, they are called 'equivalent' with regard to that property. The notion of an equivalence relation provides a means of discussing this abstractly, without referring to any particular property.

In geometry, it is often convenient to think of triangles which are congruent to one another as equivalent, even though they may be located in different places and therefore, strictly, different triangles.

In a university, it is sometimes useful to think of the students in each college or hall of residence as equivalent. Then you have divided the students of the university into subsets matched with the colleges.

In **Z**, think of all the numbers which leave the same remainder on division by 5 as equivalent. Then $0$, $\pm 5$, $\pm 10$, ... are equivalent to each other, so are $..., -9, -4, 1, 6, 11, ...$, and there are other such sets. In fact, this sense of equivalence leads you to the sets in $\mathbf{Z}_5$.

## 16.2 EQUIVALENCE RELATIONS

**Definition**: An **equivalence relation** on a set $A$ is a relation, denoted by $\sim$, between the elements of $A$ with the following properties.

- $x \sim x$ for all $x \in A$          Reflexive property
- if $x \sim y$, then $y \sim x$          Symmetric property
- if $x \sim y$ and $y \sim z$, then $x \sim z$     Transitive property

Here are some examples of equivalence relations and some which are not.

### ■ *Example 16.2.1*

Let $A = \{$people living in the UK$\}$ and let $x \sim y$ if $x$ and $y$ were born in the same calendar year.

Clearly $x$ was born in the same year as $x$, so $x \sim x$ and $\sim$ is reflexive.

If $x$ was born in the same year as $y$, then $y$ was born in the same year as $x$. So $\sim$ is symmetric.

And if $x$ was born in the same year as $y$, and $y$ was born in the same year as $z$, then $x$ was born in the same year as $z$. So $\sim$ is transitive.

Therefore $\sim$ is an equivalence relation on $A$. Notice that $A$ has been divided into subsets of people all born in the same calendar year.

### ■ *Example 16.2.2*

Show that the relation $\sim$ on **Z** defined by $x \sim y$ if $x - y$ is divisible by 5 is an equivalence relation.

Since $x - x = 0$ which is divisible by 5, $x \sim x$. So $\sim$ is reflexive.

If $x \sim y$, then $x - y$ is divisible by 5. Therefore $y - x$ is divisible by 5, so $y \sim x$. So $\sim$ is symmetric.

Finally if $x \sim y$ and $y \sim z$, then both $x - y$ and $y - z$ are divisible by 5. Therefore $x - y = 5m$ and $y - z = 5n$ for some integers $m$ and $n$. Adding these two equations gives $x - z = 5(m + n)$, so $x - z$ is divisible by 5. So $x \sim z$ and $\sim$ is transitive.

Hence $\sim$ is an equivalence relation.

Notice that $\mathbf{Z}$ has been divided up into the subsets of elements which are all related to each other. Each of these subsets is an element of $\mathbf{Z}_5$.

## ■ *Example 16.2.3*

The relation on $A = \{\text{people living in the UK}\}$, given by $x \sim y$ if $x$ is a friend of $y$, is not an equivalence relation, because you cannot guarantee that the transitive relation holds. $x$ can be friendly with $y$, and $y$ can be friendly with $z$ without $x$ being friendly with $z$.

## ■ *Example 16.2.4*

Define the relation $\sim$ on $\mathbf{Z}$, by $x \sim y$ if 5 divides $2x - y$.

This is not an equivalence relation because $x$ is not related to $x$ for all $x \in \mathbf{Z}$. If $x = 1$, then $2x - x = x = 1$, and 5 does not divide 1.

> *In fact, this relation is not symmetric or transitive, as well as not being reflexive, but you only need to show that one of the conditions fails.*

In Examples 16.2.1 and 16.2.2, in which the relations were equivalence relations, the underlying sets, $A$ and $\mathbf{Z}$, were divided into subsets containing elements which were related to each other. This idea leads to the following definition and theorem.

**Definition**: The set $\bar{a} = \{x \in A : x \sim a\}$ is called the **equivalence class** of $a$.

**Theorem 55**: Let $\sim$ be an equivalence relation on a set $A$. Then
(1) for each $a \in A$, $a \in \bar{a}$;
(2) $a \sim b$ if, and only if, $\bar{a} = \bar{b}$.

*This theorem says that a is in its own equivalence class, and if a is related to b, then the equivalence classes of a and b are identical, and vice versa.*

**Proof**: (1) Since ~ is an equivalence relation, $a \sim a$, so $a \in \bar{a}$.

(2) *If*. Suppose that $\bar{a} = \bar{b}$. Then $\bar{a} \subseteq \bar{b}$. From (i), $a \in \bar{a}$, so $a \in \bar{b}$. Therefore $a \sim b$.

*Only if*. Let $a \sim b$. First suppose that $x \in \bar{a}$, so that $x \sim a$. Then $x \sim a$ and $a \sim b$, so, by the transitive rule $x \sim b$. Therefore $x \in \bar{b}$. So if $x \in \bar{a}$, $x \in \bar{b}$. So $\bar{a} \subseteq \bar{b}$. Now suppose that $x \in \bar{b}$, so that $x \sim b$. But, $a \sim b$, so, by the symmetric rule $b \sim a$. Therefore $x \sim b$ and $b \sim a$, so, by the transitive rule, $x \sim a$. Therefore $x \in \bar{a}$. So if $x \in \bar{b}$, $x \in \bar{a}$, so $\bar{b} \subseteq \bar{a}$. Therefore, as $\bar{a} \subseteq \bar{b}$ and $\bar{b} \subseteq \bar{a}$, $\bar{a} = \bar{b}$. ∎

In Examples 16.2.1 and 16.2.2, the subsets into which $A$ and $\mathbf{Z}$ were divided were disjoint. This leads to the idea of a partition.

## 16.3 PARTITIONS

**Definition**: A **partition** of a set $A$ is a division of $A$ into subsets such that every element of $A$ is in exactly one of the subsets.

The 'exactly' part of the definition ensures that the subsets in a partition are disjoint.

**Theorem 56**: For any equivalence relation on a set $A$, the set of equivalence classes forms a partition of $A$.

**Proof**: Every element of $A$ is in at least one equivalence class by part (1) of Theorem 55.

It remains to prove that each element is in exactly one equivalence class by showing that distinct equivalence classes are disjoint. Let $a, b \in A$ and suppose that $\bar{a} \cap \bar{b}$ is non-empty. Then there exists an element $x \in A$, such that $x \in \bar{a}$ and $x \in \bar{b}$. Therefore $x \sim a$ and $x \sim b$. Therefore, as ~ is symmetric, $a \sim x$ and $x \sim b$, so, by the transitive rule, $a \sim b$. Therefore, by part (2) of Theorem 55, $\bar{a} = \bar{b}$. Therefore either $\bar{a}$ and $\bar{b}$ are identical or they are disjoint. ∎

## ■ *Example 16.3.1*

Returning to Example 16.2.1 in which the equivalence relation was $x \sim y$ if $x$ and $y$ were born in the same calendar year, the equivalence classes are the sets of people all born in the same calendar year.

## ■ *Example 16.3.2*

Returning to Example 16.2.2, where, on $\mathbf{Z}$, the equivalence relation was $x \sim y$ if $x - y$ is divisible by 5, the equivalence classes are the elements of $\mathbf{Z}_5$. In fact, this is another way of looking at $\mathbf{Z}_5$.

## ■ *Example 16.3.3*

Let $H$ be a subgroup of a group $G$, and define the relation $\sim$ on $G$ by $a \sim b$ if $a^{-1}b \in H$.

For all $a \in H$, $a^{-1}a = e \in H$. Therefore $a \sim a$, so $\sim$ is reflexive.

If $a \sim b$, then $a^{-1}b \in H$, so, as $H$ is a subgroup, $\left(a^{-1}b\right)^{-1} \in H$. Therefore $b^{-1}a \in H$, so $b \sim a$. So $\sim$ is symmetric.

Finally, if $a \sim b$ and $b \sim c$, then $a^{-1}b \in H$ and $b^{-1}c \in H$. Therefore, as $H$ is a subgroup, $\left(a^{-1}b\right)\left(b^{-1}c\right) \in H$ or $a^{-1}c \in H$. Therefore $a \sim c$. So $\sim$ is transitive.

Therefore $\sim$ is an equivalence relation on $G$.

The equivalence class containing $a$ is $\bar{a} = \{x \in G : x \sim a\}$, that is $\bar{a} = \left\{x \in G : x^{-1}a \in H\right\}$.

*What is going on is that there are two partitions on G, the cosets of H and the equivalence classes of ~. Theorem 46 says that a and b are in the same coset if, and only if, $aH = bH$; and that $aH = bH$ if, and only if, $a^{-1}b \in H$. Theorem 55 says that a and b belong to the same equivalence class of ~ if, and only if, $\bar{a} = \bar{b}$; and that $\bar{a} = \bar{b}$ if, and only if, $a \sim b$. Example 16.3.3 links these two theorems because it says that $a \sim b$ if, and only if, $a^{-1}b \in H$. Therefore a and b are in the same coset if, and only if, a and b are in the same equivalence class of ~. So the equivalence classes of the relation ~ on G are precisely the left cosets of H in G.*

You can give an explicit proof of the fact that $\bar{a} = aH$. If $x \in \bar{a}$, then $a^{-1}x \in H$. From Theorem 46, part (1), $aH = xH$, so, from Theorem 46, part (2), $x \in aH$. If $x \in aH$, then, from Theorem 46, part (2), $aH = xH$. By Theorem 46, part (1), $a^{-1}x \in H$, so $x \in \bar{a}$. Therefore $\bar{a} = aH$.

### ■ *Example 16.3.4*

Here is a non-numerical example. $A$ = the set of towns in Great Britain. Let $a$ and $b$ be towns, and let $a \sim b$ if $a$ and $b$ are in the same county. Then $\sim$ is reflexive, symmetric and transitive, and the equivalence classes are the counties. Fig. 16.1 shows Great Britain partitioned into counties.

**Fig. 16.1** *Great Britain partitioned into equivalence classes*

## 16.4 AN IMPORTANT EQUIVALENCE RELATION

Here is an example which will be used in Theorem 64 in Chapter 19.

Let $f : X \to Y$ be any function. Then $a \sim b$ if $f(a) = f(b)$ defines an equivalence relation on $X$.

As $f(a) = f(a)$, $a \sim a$ and $\sim$ is reflexive.

If $a \sim b$, then $f(a) = f(b)$. It follows that $f(b) = f(a)$, so $b \sim a$ and $\sim$ is symmetric.

Finally, if $a \sim b$ and $b \sim c$, then $f(a) = f(b)$ and $f(b) = f(c)$, so $f(a) = f(c)$ and $a \sim c$, so $\sim$ is symmetric. Therefore $\sim$ is an equivalence relation on $X$. The equivalence classes are called the **fibres** of $f$. The fibres of $f$ are illustrated in Fig. 16.2.

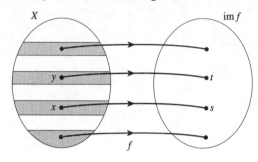

**Fig. 16.2** *The fibres of a function f*

The fibres of $f$ are the horizontal regions of $X$ containing all the elements which map on to each single element in the image of $f$, im $f$.

Fig. 16.2 suggests that there is a bijection from the fibres of $X$ to the image of $f$. Call the set of fibres $X/\sim$.

**Theorem 57**: Let $f : X \to Y$ be any function, and let $\sim$ be the equivalence relation defined by $a \sim b$ if $f(a) = f(b)$. Then the function $\theta : X/\sim \to$ im $f$ defined by $\theta(\bar{x}) = f(x)$ is a bijection.

**Proof**:

*The proof has three parts. First, the function $\theta$ must be shown to be well-defined, that is, if $\bar{x} = \bar{y}$, then $\theta(\bar{x}) = \theta(\bar{y})$. The other two parts are the normal surjection and injection parts of proving that a function is a bijection.*

*Well defined.* Suppose that $\bar{x} = \bar{y}$. Then, by Theorem 55, $x \sim y$, so $f(x) = f(y)$. Therefore $\theta$ is well defined.

*Injection.* Suppose that $\theta(\bar{x}) = \theta(\bar{y})$. Then $f(x) = f(y)$, and therefore $x \sim y$. Therefore, by Theorem 55, $\bar{x} = \bar{y}$, so $\theta$ is injective.

*Surjection.* For each $y$ in the image of $f$ there exists an $x \in X$ such that $f(x) = y$. For such an $x$, there is an equivalence class $\bar{x}$, and for that $\bar{x}$, $\theta(\bar{x}) = f(x) = y$. Therefore $\theta$ is surjective.

Therefore $\theta$ is a bijection ∎.

## WHAT YOU SHOULD KNOW

- The definition of an equivalence relation.

- The meaning of 'reflexive', 'symmetric' and 'transitive' in the context of relations.

- The definition of a partition.

- That an equivalence relation on a set partitions the set into equivalence classes.

## EXERCISE 16

**1** Show that each of the following relations is an equivalence relation. In each case identify the equivalence classes.

(1)  On $\mathbf{R}^2 - (0,0)$, $(a,b) \sim (c,d)$ if $ad - bc = 0$

(2)  On $\mathbf{Q}$, $p/q \sim r/s$ if $ps - qr = 0$

(3)  On $\mathbf{Z}$, $x \sim y$ if $x - y$ is divisible by 2

(4)  On $\mathbf{Z}$, $x \sim y$ if $2x + y$ is divisible by 3

(5)  On $\mathbf{Z}$, $x \sim y$ if $|x| = |y|$

**2** Decide whether each of the following relations is an equivalence relation, giving the equivalence classes where appropriate.

(1)  On $\mathbf{Z}$, $x \sim y$ if $x - y$ is the square of an integer

(2)  On $\mathbf{Z}$, $x \sim y$ if $xy > 0$

(3)  On $\mathbf{Z}^+$, $x \sim y$ if $xy > 0$

(4)  On $\mathbf{Z}$, $x \sim y$ if $xy \geq 0$

(5)  On $\mathbf{R}$, $a \sim b$ if $|a - b| \leq \frac{1}{2}$

(6)  On $\mathbf{R}$, $a \sim b$ if $z \in \mathbf{Z}$ exists so that $|z - a| \leq \frac{1}{2}$ and $|z - b| \leq \frac{1}{2}$

(7)  On the set of lines in a plane, $l \sim m$ if $l$ is parallel to $m$

(8)  On $\mathbf{R}^2$ $(a,b) \sim (c,d)$ if $bd = 0$

(9)  On the set of triangles, $A \sim B$ if $A$ is similar to $B$

(10)  On the set of triangles, $A \sim B$ if $A$ is congruent to $B$

(11)  On the set of lines in a plane, $l \sim m$ if $l$ is perpendicular to $m$

# 17

## *Quotient groups*

### 17.1 INTRODUCTION

In Chapter 8 you saw how new, bigger groups could be built up from smaller groups by using the Cartesian product. The multiplication sign × was used to denote the Cartesian product.

You probably have wondered whether there is any process which corresponds in some way to division. The answer is yes – under some circumstances. The purpose of this chapter is to investigate those circumstances.

Figure 17.1 shows the group $D_3$, with one difference – the subgroup $H = \{e, a, a^2\}$ is shown shaded and the coset $bH$ is shown unshaded. If you hold the page with Fig. 17.1 sufficiently far away so that you cannot see the details of the individual elements, what you see is shown in Fig. 17.2. This is also the table of a group in which there are two elements, the colours grey and white. This group of colours is isomorphic to $\mathbf{Z}_2$.

The way that this situation is described is to use a form of division notation and to write $D_3/H \cong \mathbf{Z}_2$. You might also observe that as $H$ is isomorphic to $\mathbf{Z}_3$, you could write $D_3/\mathbf{Z}_3 \cong \mathbf{Z}_2$. $D_3/\mathbf{Z}_3$ is an example of a quotient group, or a factor group.

|       | $e$     | $a$     | $a^2$   | $b$     | $ba$    | $ba^2$  |
|-------|---------|---------|---------|---------|---------|---------|
| $e$   | $e$     | $a$     | $a^2$   | $b$     | $ba$    | $ba^2$  |
| $a$   | $a$     | $a^2$   | $e$     | $ba^2$  | $b$     | $ba$    |
| $a^2$ | $a^2$   | $e$     | $a$     | $ba$    | $ba^2$  | $b$     |
| $b$   | $b$     | $ba$    | $ba^2$  | $e$     | $a$     | $a^2$   |
| $ba$  | $ba$    | $ba^2$  | $b$     | $a^2$   | $e$     | $a$     |
| $ba^2$| $ba^2$  | $b$     | $ba$    | $a$     | $a^2$   | $e$     |

**Fig. 17.1** *The group* $D_3$

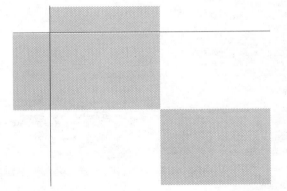

**Fig. 17.2** *The group* $D_3/\mathbf{Z}_3 \cong \mathbf{Z}_2$

However, you cannot deduce from $D_3/\mathbf{Z}_3 \cong \mathbf{Z}_2$ that $D_3 \cong \mathbf{Z}_2 \times \mathbf{Z}_3$. (You saw in Section 11.5 that $\mathbf{Z}_2 \times \mathbf{Z}_3 \cong \mathbf{Z}_6$, and in Section 15.4 that $D_3$ is not isomorphic to $\mathbf{Z}_6$.)

Here is another example, this time from $D_4$, starting with the subgroup $H = \left\{ e, a^2 \right\}$. Figure 17.3 shows $D_4$ with the subsets shaded differently from each other.

This time if you hold Fig. 17.3 so that you cannot see the details of the individual elements you see the group $V$, consisting of four elements.

You can write $D_4/H = V$, or, as $H$ has two elements it is isomorphic to $\mathbf{Z}_2$, so $D_4/\mathbf{Z}_2 = V$.

|        | $e$    | $a^2$  | $a$    | $a^3$  | $b$    | $ba^2$ | $ba$   | $ba^3$ |
|--------|--------|--------|--------|--------|--------|--------|--------|--------|
| $e$    | $e$    | $a^2$  | $a$    | $a^3$  | $b$    | $ba^2$ | $ba$   | $ba^3$ |
| $a^2$  | $a^2$  | $e$    | $a^3$  | $a$    | $ba^2$ | $b$    | $ba^3$ | $ba$   |
| $a$    | $a$    | $a^3$  | $a^2$  | $e$    | $ba^3$ | $ba$   | $b$    | $ba^2$ |
| $a^3$  | $a^3$  | $a$    | $e$    | $a^2$  | $ba$   | $ba^3$ | $ba^2$ | $b$    |
| $b$    | $b$    | $ba^2$ | $ba$   | $ba^3$ | $e$    | $a^2$  | $a$    | $a^3$  |
| $ba^2$ | $ba^2$ | $b$    | $ba^3$ | $ba$   | $a^2$  | $e$    | $a^3$  | $a$    |
| $ba$   | $ba$   | $ba^3$ | $ba^2$ | $b$    | $a^3$  | $a$    | $e$    | $a^2$  |
| $ba^3$ | $ba^3$ | $ba$   | $b$    | $ba^2$ | $a$    | $a^3$  | $a^2$  | $e$    |

**Fig. 17.3**

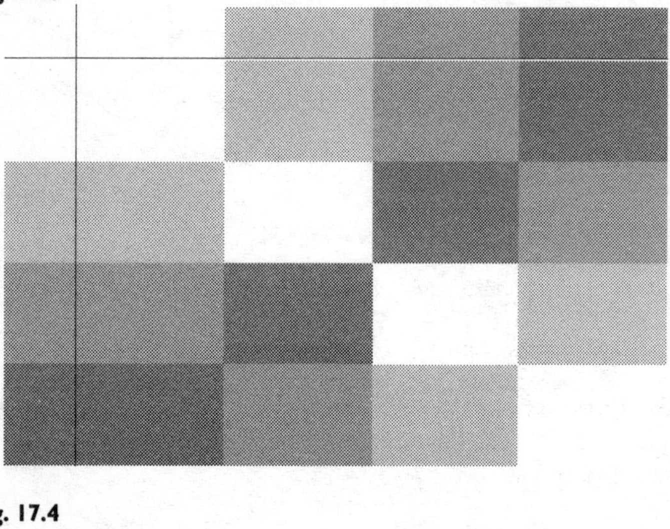

**Fig. 17.4**

Here is another example. This time the group $D_3$ has been rearranged so that the subgroup $H = \{e, b\}$, in white, is in the top left-hand corner of Fig. 17.5. The left coset $aH = \{a, ab\}$ is shown lightly shaded, and the third coset, $a^2H = \{a^2, a^2b\}$, is shown in a darker shading.

You can see from the shading in Fig. 17.5 that there is no longer a tidy pattern. It is not now possible to view Fig. 17.5 at a distance and see the colours form the structure of a group. In this case, you cannot write $D_3/\mathbf{Z}_2 \cong \mathbf{Z}_3$.

*Remember that $ab = ba^2$ and $a^2b = ba$.*

|       | $e$    | $b$    | $a$    | $ab$   | $a^2$   | $a^2b$  |
|-------|--------|--------|--------|--------|---------|---------|
| $e$   | $e$    | $b$    | $a$    | $ab$   | $a^2$   | $a^2b$  |
| $b$   | $b$    | $e$    | $a^2b$ | $a^2$  | $ab$    | $a$     |
| $a$   | $a$    | $ab$   | $a^2$  | $a^2b$ | $e$     | $b$     |
| $ab$  | $ab$   | $a$    | $b$    | $e$    | $a^2b$  | $a^2$   |
| $a^2$ | $a^2$  | $a^2b$ | $e$    | $b$    | $a$     | $ab$    |
| $a^2b$| $a^2b$ | $a^2$  | $ab$   | $a$    | $b$     | $e$     |

**Fig. 17.5**

So what are the characteristics of the subgroups and cosets which enable you to say that $D_3/\mathbf{Z}_3 \cong \mathbf{Z}_2$, but not $D_3/\mathbf{Z}_2 \cong \mathbf{Z}_3$ or $D_3 \cong \mathbf{Z}_2 \times \mathbf{Z}_3$?

## 17.2 SETS AS ELEMENTS OF SETS

This section heading is not a misprint; it really is about sets of which the elements are sets. You have already met some examples: for instance, $\mathbf{Z}_n$ in Chapter 4, and the set of fibres $X/\sim$ of a function in Chapter 16. Here are some more examples.

## ■ *Example 17.2.1*

Consider the elements of $\mathbf{Z}_{10}$. You are familiar with using the elements of $\mathbf{Z}_{10}$ to check arithmetic results. For example, you know immediately that $2349 \times 3487 \neq 8\,910\,961$ because the units digit is wrong. What you are actually doing, probably implicitly rather than explicitly, is the following process.

The original numbers $2349 \in [9]_{10}$ and $3487 \in [7]_{10}$. Multiplying $[9]_{10}$ by $[7]_{10}$ gives $[9]_{10} \times [7]_{10} = [3]_{10}$. Finally $8\,910\,961 \notin [3]_{10}$ so the result of the multiplication is wrong.

Remember that a statement like $[9]_{10} \times [7]_{10} = [3]_{10}$ means that *if you take any element from the set* $[9]_{10}$, *and multiply it by any element from the set* $[7]_{10}$, *you obtain a result which is in the set* $[3]_{10}$.

## ■ *Example 17.2.2*

Suppose now that you were to devise a process called digital product. The digital product of a non-negative integer is found by repeatedly multiplying the digits together.

The sets in the partition of the non-negative integers are the following, where the first 30 numbers, together with some others are shown:

$\langle 0 \rangle = \{0, 10, 20, 25, 30, \ldots, 56, \ldots, 69, \ldots\}$

$\langle 1 \rangle = \{1, 11, \ldots\}$

$\langle 2 \rangle = \{2, 12, 21, 26, \ldots\}$

$\langle 3 \rangle = \{3, 13, \ldots\}$

$\langle 4 \rangle = \{4, 14, 22, 27, \ldots\}$

$\langle 5 \rangle = \{5, 15, \ldots\}$

$\langle 6 \rangle = \{6, 16, 23, 28, \ldots, 48, \ldots, 84, \ldots\}$

$\langle 7 \rangle = \{7, 17, \ldots\}$

$\langle 8 \rangle = \{8, 18, 24, 29, \ldots\}$

$\langle 9 \rangle = \{9, 19, \ldots\}.$

Now look at the product $6 \times 3 = 18$. The digital product of 18 is 8 so $18 \in \langle 8 \rangle$. But if you look at $\langle 6 \rangle \times \langle 3 \rangle$ by taking *any possible product of elements* in $\langle 6 \rangle$ and $\langle 3 \rangle$ you get $\{18, 48, 56, 69, 84, ...\}$. But there is no single set which has these numbers as elements. In this case you cannot multiply the sets in a way which reflects the multiplication of the integers.

**Definition**: Let $X$ and $Y$ be subsets of a set $A$ and let a binary operation $\circ$ be defined on $A$. Then the set $XY =$ the set of all elements of $A$ which can be expressed in the form $x \circ y$, where $x \in X$, $y \in Y$. This operation on the set of subsets of $A$ is called the **operation induced by the operation** $\circ$ **in** $A$.

Notice that the products in $XY$ need not be distinct; you could have $x_1 \neq x_2$ and $y_1 \neq y_2$ but $x_1 y_1 = x_2 y_2$. However, repetitions are ignored. Moreover, two subsets are equal if, and only if, they have the same distinct elements, irrespective of repetitions.

You are advised to go to Exercises 17, and try questions 1 and 2.

## 17.3 COSETS AS ELEMENTS OF A GROUP

So what has all this got to do with cosets?

Let $H$ be a subgroup of a group $G$, and let $a$ and $b$ be any two elements of $G$. Suppose also that the cosets of $H$ form a group under the operation on the set of subsets of $G$ induced by the operation in $G$. The two cosets corresponding to $a$ and $b$ are $aH$ and $bH$. When you work out the product of these two cosets to get $(aH)(bH)$ you must take every element in $aH$ and multiply it by every element in $bH$.

As the set of cosets is a group, it must be closed under the operation and so the resultant product $(aH)(bH)$ must be another coset. Which coset will it be? Since $e \in H$, $a = ae \in aH$ and $b = be \in bH$, so one element in the set $(aH)(bH)$ is $ab$. So the coset of the product $(aH)(bH)$ must contain $ab$, and therefore, by Theorem 45, part (2), it must be $(ab)H$.

If for all $a$ and $b$ in $G$ the product $(aH)(bH) = (ab)H$, what can you say about the subgroup $H$? The answer is encapsulated in Theorem 58.

**Theorem 58**: Let $H$ be a subgroup of a group $G$. Then $(aH)(bH) = (ab)H$ for all $a$ and $b$ in $G$ if, and only if, $x^{-1}Hx \subseteq H$ for all $x \in G$.

*If.* Suppose that $x^{-1}Hx \subseteq H$ for all $x \in G$. Let $a, b \in G$.

> *You need to show first that $(aH)(bH) \subseteq (ab)H$.*

Consider $g \in (aH)(bH)$. Then $g = (ah_1)(bh_2)$ where $h_1, h_2 \in H$. So $g = ab(b^{-1}h_1b)h_2$. Now $x^{-1}Hx \subseteq H$ for all $x \in G$, so $b^{-1}h_1b = h$ for some $h \in H$. Thus $g = abhh_2$. As $H$ is a subgroup, $hh_2 \in H$, so $g = abh_3$ where $h_3 \in H$, so $g \in (ab)H$. Therefore $(aH)(bH) \subseteq (ab)H$.

> *Now show that $(ab)H \subseteq (aH)(bH)$.*

Consider $g \in (ab)H$. Then $g = abh_1$, where $h_1 \in H$, and therefore $g = a(bh_1b^{-1})b$. Now $x^{-1}Hx \subseteq H$ for all $x \in G$, so, with $x = b^{-1}$, $bh_1b^{-1} = h$ for some $h \in H$. Thus $g = ahb = (ah)(be)$, so $g \in (aH)(bH)$. Therefore $(ab)H \subseteq (aH)(bH)$.

Therefore, as $(aH)(bH) \subseteq (ab)H$ and $(ab)H \subseteq (aH)(bH)$, it follows that $(aH)(bH) = (ab)H$.

*Only if.* Suppose that $(aH)(bH) = (ab)H$ for all $a$ and $b$ in $G$ and let $x$ be an element of $G$. If $g \in x^{-1}Hx$, then $g = x^{-1}hx$ for some $h \in H$. So $g = (x^{-1}h)(xe)$ which is a member of $(x^{-1}H)(xH)$, and therefore of $(x^{-1}x)H$ or $eH = H$. Therefore $x^{-1}Hx \subseteq H$. ∎

## 17.4 NORMAL SUBGROUPS

The condition that $x^{-1}Hx \subseteq H$ for all $x \in G$ is important, and will give rise to a definition. But first notice that it implies $x^{-1}Hx = H$ for all $x \in G$.

**Theorem 59**: Let $x^{-1}Hx \subseteq H$ for all $x \in G$. Then $x^{-1}Hx = H$ for all $x \in G$.

**Proof**: Let $x$ be any element of $G$. You have in this case to prove only that $H \subseteq x^{-1}Hx$. Take any element $h$ of $H$. Then $h = x^{-1}(xhx^{-1})x$ $= x^{-1}kx$ where $k = xhx^{-1}$. But $k = y^{-1}hy$ where $y = x^{-1}$. Therefore, as $x^{-1}Hx \subseteq H$ for all $x \in G$ (in particular $y^{-1}Hy \subseteq H$), $k \in H$. Hence $h = x^{-1}kx \in x^{-1}Hx$. Therefore $H \subseteq x^{-1}Hx$, so $x^{-1}Hx = H$. ∎

**Definition:** A subgroup $H$ of a group $G$ is a **normal subgroup** if $x^{-1}Hx = H$ for each $x \in G$.

> *Theorem 59 tells us that to show that a subgroup $H$ is normal it is sufficient to show that $x^{-1}Hx \subseteq H$ for all $x \in G$.*
>
> *For an abelian group $G$, every subgroup is normal. This is because, for all $x \in G$ and all $h \in H$, $x^{-1}hx = hx^{-1}x = he = h \in H$.*

Let $H$ be a subgroup of $G$. Can you say that the cosets of $H$ form a group, with the multiplication induced from $G$, if $H$ is a normal subgroup? The answer is yes.

From the discussion at the beginning of Section 17.3 it follows that the set of cosets is closed under the operation (on the set of subsets of $G$ induced by the operation in $G$) if, and only if, $(aH)(bH) = (ab)H$ for all $a$ and $b$ in $G$.

Theorem 58 shows that $(aH)(bH) = (ab)H$ for all $a$ and $b$ in $G$ if, and only if, $H$ is a normal subgroup of $G$.

Combining these two statements, the normality of $H$ is a necessary and sufficient condition for the set of cosets of $H$ to be closed under the operation of multiplication of subsets induced by the operation of $G$.

We have gone a long way towards the next theorem.

**Theorem 60**: Let $G$ be a group, and let $H$ be a normal subgroup. Then the set of cosets of $H$ forms a group under the operation on the set of subsets of $G$ induced by the group operation of $G$.

**Proof**: You already know from the remarks before the theorem that the closure axiom is satisfied.

To prove that the operation of coset multiplication is associative, note that $aH((bH)(cH)) = aH((bc)H) = (a(bc))H$. Similarly, you find that $((aH)(bH))cH = ((ab)H)cH = ((ab)c)H$. But the group operation in $G$ is associative, that is, $a(bc) = (ab)c$, so $(a(bc))H = ((ab)c)H$, and the result follows.

The coset $eH$ (or $H$) is the identity, because $(eH)(aH) = (ea)H = aH$ and $(aH)(eH) = (ae)H = aH$.

Finally $a^{-1}H$ is the inverse of $aH$ as $(a^{-1}H)(aH) = (a^{-1}a)H = eH$ and $(aH)(a^{-1}H) = (aa^{-1})H = eH$.

All four group axioms are satisfied so the cosets of $H$ form a group under the operation on the set of subsets of $G$ induced by the group operation of $G$. ∎

**Definition**: Let $H$ be a normal subgroup of $G$. Then the group of cosets featuring in Theorem 60 is called the **quotient group** of $H$ in $G$, and written $G/H$. The term 'factor group' is sometimes used to mean quotient group.

Note that if $G$ is finite, then $G/H$ is finite. $G$ is a disjoint union of cosets and, as established in the proof of Lagrange's theorem, the number of elements in each of these cosets is equal to the order of $H$. Therefore the number of elements in $G$ is equal to the number of cosets multiplied by the number of elements in $H$. Therefore the order of the group $G/H$ is the order of $G$, divided by the order of $H$.

## 17.5 THE QUOTIENT GROUP

Here are some examples of quotient groups.

### ■ *Example 17.5.1*

In $D_6$, show that the subgroup $H = \{e, a^3\}$ is normal. Find the group $D_6/H$.

> *To prove that $H$ is normal, you have to prove that $g^{-1}Hg \subseteq H$ for all $g \in G$.*

$D_6$ is generated by $a$ and $b$ such that $a^6 = b^2 = e$ and $aba = b$. Consider elements of the form $g^{-1}Hg$. These are of two types, $g^{-1}eg$ and $g^{-1}a^3g$. For the first case $g^{-1}eg = g^{-1}g = e \in H$. Suppose that for the second case, $g = b^i a^j$, where $i = 0$ or $1$, and $j = 0, 1, 2, 3, 4$ or $5$. Then since $\left(b^i\right)^{-1} = b^i$, $g^{-1}a^3g = \left(b^i a^j\right)^{-1} a^3 b^i a^j = a^{-j} b^i a^3 b^i a^j$. But if $i = 0$, $b^i a^3 b^i = a^3$. If $i = 1$, $b^i a^3 b^i = ba^3 b = (bab)(bab)(bab)$. But as $aba = b$, $bab = a^{-1}(aba)b = a^{-1}bb = a^{-1}$, and $ba^3 b = a^{-3} = a^3 \in H$. Thus $g^{-1}Hg = H$ for all $g \in G$.

Thus $H$ is a normal subgroup, so $D_6/H$ is a group. From the remark at the end of the previous section, this group has $24 \div 4 = 6$ elements. The elements of $D_6$ are $e, a, a^2, a^3, a^4, a^5, b, ba, ba^2, ba^3, ba^4$ and $ba^5$. As $a^3 \in H$, it follows from Theorem 46 that every left coset of $H$ in $D_6$ is one of $H, aH, a^2H, bH, baH$ or $ba^2H$. So the quotient group $D_6/H$ consists of these six cosets.

A typical calculation such as $(a^2H)(bH)$ in $D_6/H$ is carried out by first saying $(a^2H)(bH) = (a^2b)H$. Then in $D_6$, using the result $a^iba^i = b$ from Section 13.3, $a^2b = (a^2ba^2)a^{-2} = ba^{-2} = ba^4$, so $(a^2b)H = (ba^4)H = baH$. Therefore $(a^2H)(bH) = baH$.

Figure 17.6 shows its group table.

|        | $H$      | $aH$     | $a^2H$   | $bH$     | $baH$    | $ba^2H$  |
|--------|----------|----------|----------|----------|----------|----------|
| $H$    | $H$      | $aH$     | $a^2H$   | $bH$     | $baH$    | $ba^2H$  |
| $aH$   | $aH$     | $a^2H$   | $H$      | $ba^2H$  | $bH$     | $baH$    |
| $a^2H$ | $a^2H$   | $H$      | $aH$     | $baH$    | $ba^2H$  | $bH$     |
| $bH$   | $bH$     | $baH$    | $ba^2H$  | $H$      | $aH$     | $a^2H$   |
| $baH$  | $baH$    | $ba^2H$  | $bH$     | $a^2H$   | $H$      | $aH$     |
| $ba^2H$| $ba^2H$  | $bH$     | $baH$    | $aH$     | $a^2H$   | $H$      |

**Fig. 17.6** *The group* $D_6/H$

If you compare the group table in Fig. 17.6 with the table for $D_3$ in Example 5.2.2 you will see that they are the same, with:

$$\begin{array}{cccccc} H & aH & a^2H & bH & baH & ba^2H \\ \updownarrow & \updownarrow & \updownarrow & \updownarrow & \updownarrow & \updownarrow \\ I & R & S & X & Y & Z \end{array}$$

and this shows that $D_6/H$ is isomorphic to $D_3$.

## ■ *Example 17.5.2*

Consider the group $(\mathbf{Z}, +)$, and the subgroup $(n\mathbf{Z}, +)$ consisting of the multiples of $n$. This subgroup is normal, because the group $(\mathbf{Z}, +)$ is abelian, so the condition for normality is automatically satisfied.

The cosets of $n\mathbf{Z}$ are $n\mathbf{Z}$, $1 + n\mathbf{Z}$, $2 + n\mathbf{Z}$, ... , $(n-1) + n\mathbf{Z}$. This set of cosets forms a group under the operation given by $(r + n\mathbf{Z}) + (s + n\mathbf{Z}) = (r + s) + n\mathbf{Z}$. In fact, it is not difficult to see that $\mathbf{Z}/n\mathbf{Z} \cong \mathbf{Z}_n$.

## ■ *Example 17.5.3*

Determine the group $(\mathbf{Z}_4 \times \mathbf{Z}_8) / (\mathbf{Z}_4 \times \{0\})$.

The group $\mathbf{Z}_4 \times \mathbf{Z}_8$ has 32 elements and the subgroup $\mathbf{Z}_4 \times \{0\}$ has four elements. Therefore $\mathbf{Z}_4 \times \{0\}$ has eight cosets, so the quotient group $(\mathbf{Z}_4 \times \mathbf{Z}_8) / (\mathbf{Z}_4 \times \{0\})$ has eight elements.

Each element of $\mathbf{Z}_4 \times \mathbf{Z}_8$ is of the form $(a, b)$ where $a$ is one of the four residue classes $[i]_4$, $0 \le i \le 3$, and $b$ is one of the eight residue classes $[j]_8$, $0 \le j \le 7$. But $(a, b) - (0, b) = (a, 0) \in \mathbf{Z}_4 \times \{0\}$. Therefore, by Theorem 46, every coset of $\mathbf{Z}_4 \times \{0\}$ is of the form $(0, b) + (\mathbf{Z}_4 \times \{0\})$. Therefore the elements of the quotient group $(\mathbf{Z}_4 \times \mathbf{Z}_8) / (\mathbf{Z}_4 \times \{0\})$ are the eight cosets $(0, 0) + (\mathbf{Z}_4 \times \{0\})$, $(0, 1) + (\mathbf{Z}_4 \times \{0\})$, $(0, 2) + (\mathbf{Z}_4 \times \{0\})$, $(0, 3) + (\mathbf{Z}_4 \times \{0\})$, ... , $(0, 7) + (\mathbf{Z}_4 \times \{0\})$.

The order of the element $(0, 1) + (\mathbf{Z}_4 \times \{0\})$ is 8, so $(\mathbf{Z}_4 \times \mathbf{Z}_8) / (\mathbf{Z}_4 \times \{0\})$ is the cyclic group $\mathbf{Z}_8$. Thus $(\mathbf{Z}_4 \times \mathbf{Z}_8) / (\mathbf{Z}_4 \times \{0\}) \cong \mathbf{Z}_8$.

*Notice that it is as though the group $\mathbf{Z}_4$ has cancelled leaving $\mathbf{Z}_8$.*

## ■ *Example 17.5.4*

Determine the group $(\mathbf{Z}_8 \times \mathbf{Z}_4) / \langle (2, 0) \rangle$.

The group $\mathbf{Z}_8 \times \mathbf{Z}_4$ has 32 elements and the subgroup $\langle (2, 0) \rangle$ generated by $(2, 0)$ has four elements. So $\langle (2, 0) \rangle$ has eight cosets in $\mathbf{Z}_8 \times \mathbf{Z}_4$, so the quotient group $(\mathbf{Z}_8 \times \mathbf{Z}_4) / \langle (2, 0) \rangle$ has eight elements.

Each element of $\mathbf{Z}_8 \times \mathbf{Z}_4$ is of the form $(a,b)$ where $a$ is one of the eight residue classes $[i]_8$, $0 \le i \le 7$, and $b$ is one of the four residue classes $[j]_4$, $0 \le j \le 3$. If $a$ is the class of 0, 2, 4 or 6, then $(a,b) - (0,b) = (a,0) \in \langle (2,0) \rangle$; while if $a$ is the class of 1, 3, 5 or 7, then $(a,b) - (1,b) = (a-1,0) \in \langle (2,0) \rangle$. Therefore, by Theorem 46, every coset of $\langle (2,0) \rangle$ is one of $(0,0) + \langle (2,0) \rangle$, that is, the subgroup $\langle (2,0) \rangle$ itself, $(0,1) + \langle (2,0) \rangle$, $(0,2) + \langle (2,0) \rangle$, $(0,3) + \langle (2,0) \rangle$, $(1,0) + \langle (2,0) \rangle$, $(1,1) + \langle (2,0) \rangle$, $(1,2) + \langle (2,0) \rangle$ and $(1,3) + \langle (2,0) \rangle$. Therefore these eight cosets are the elements of $(\mathbf{Z}_8 \times \mathbf{Z}_4)/\langle (2,0) \rangle$.

If you check the orders of these elements of $(\mathbf{Z}_8 \times \mathbf{Z}_4)/\langle (2,0) \rangle$, you find that there are four elements of order 4 and three elements of order 2, so, from Chapter 15, the quotient group is isomorphic to $\mathbf{Z}_2 \times \mathbf{Z}_4$. Therefore you can write $(\mathbf{Z}_8 \times \mathbf{Z}_4)/\langle (2,0) \rangle \cong \mathbf{Z}_2 \times \mathbf{Z}_4$.

## WHAT YOU SHOULD KNOW

■ The rule for multiplying together two subsets of a group.

■ How to combine two cosets using the rule induced by the operation of the group.

■ The meaning of a normal subgroup, and how to test for one.

■ The meaning of a quotient group.

## EXERCISE 17

**1** In the group $D_3$, compute the set $AB$ for the given sets $A$ and $B$.

(1) $A = \{a,b\}$, $B = \{e,ba\}$

(2) $A = \{e,ba^2\}$, $B = \{e,a,a^2\}$

(3) $A = \{a,b\}$, $B = \{e,a,a^2\}$

(4) $A = \{b,ba,ba^2\}$, $B = \{e,a,a^2\}$

**2** In the quaternion group $\mathbf{Q}_4$, shown in Section 15.5, let $H = \{e,a^2\}$, $A = \{a,ba\}$ and $B = \{ba,ba^3\}$. Compute the sets $HA$, $HB$, $AB$, $AH$, $BH$ and $BA$.

**3** In the group $D_3$, let $H$ be the subgroup $H = \{e,a,a^2\}$. List the sets $aH$ and $Hb$.

**4** Determine the group $(\mathbf{Z}_6 \times \mathbf{Z}_4)/\langle (2,2) \rangle$.

**5** Determine the group $(\mathbf{Z}_6 \times \mathbf{Z}_4)/\langle (3,2) \rangle$.

**6**   Show that the subgroup $H = \{e, b\}$ of $D_3$ is not normal.

**7**   Mark each of the following statements true or false.

(1)   Two cosets are either identical or disjoint.

(2)   No two cosets have the same number of elements.

(3)   You can combine any two cosets using the rule induced by the group, by taking any element from the first and combining it with any element from the second, and finding the coset which includes the resulting element.

**8**   Use the table in Fig. 15. 4 to prove that $H = \{e, a^2\}$ is a normal subgroup of $Q_4$. Identify the group $Q_4/H$.

**9**   Let $G$ be a group. Prove that $H$ is a normal subgroup of $G$ if, and only if, every right coset of $H$ in $G$ is also a left coset of $H$ in $G$.

# 18

# *Homomorphisms*

## 18.1 HOMOMORPHISMS

In Chapter 11, you studied isomorphisms. You saw there that two groups $(G, \circ)$, and $(H, \bullet)$ were isomorphic if there was a bijection $f : G \to H$ such that for each $x \in G$ and $y \in G$, $f(x \circ y) = f(x) \bullet f(y)$.

In this chapter you will study homomorphisms. The difference between a homomorphism and an isomorphism is that the function $f$ is no longer required to be a bijection.

**Definition**: Let $(G, \circ)$ and $(H, \bullet)$ be groups. Then a **homomorphism** is a function $f : G \to H$ such that $f(x \circ y) = f(x) \bullet f(y)$ for each $x \in G$ and $y \in G$.

> *The word 'homomorphism' means same structure, so you should expect some of the structure properties of the group G to be reflected in the image of G in H.*

Here are some examples.

## ■ *Example 18.1.1*

Consider the function from $(\mathbf{Z},+)$ to the group $(\{1,i,-1,-i\},\times)$ defined by $f(n) = i^n$.

This is a homomorphism since $f(m)f(n) = i^m i^n = i^{m+n} = f(m+n)$. The elements which map to 1 are $\{\ldots,-8,-4,0,4,8,12,\ldots\}$, those to $i$ are $\{\ldots,-7,-3,1,5,9,13,\ldots\}$, those to $-1$ are $\{\ldots,-6,-2,2,6,10,\ldots\}$ and those to $-i$ are $\{\ldots,-5,-1,3,7,11,\ldots\}$.

In fact the group $(\mathbf{Z},+)$ has been partitioned into four cosets, by the equivalence relation $x \sim y$ if $f(x) = f(y)$. Each element of a given coset maps into the same element in $(\{1,i,-1,-i\},\times)$.

## ■ *Example 18.1.2*

Let $(\mathbf{M},\times)$ be the group of invertible $2\times 2$ matrices with real entries under matrix multiplication and $(\mathbf{R}^*,\times)$ be the set of non-zero real numbers under multiplication.

> By definition, A is invertible if, and only if, there exists a matrix B such that $AB = BA = I$, that is, if, and only if, A has an inverse under multiplication. Recall that a matrix A is invertible if, and only if, $\det A \neq 0$.

Then $f:(\mathbf{M},\times) \to (\mathbf{R}^*,\times)$ where $f(A) = \det A$ is well-defined, since $\det A \neq 0$, and also a homomorphism, because $f(A)f(B) = \det A \det B = \det(AB) = f(AB)$.

The elements of $(\mathbf{M},\times)$ which map on to the identity in $(\mathbf{R}^*,\times)$ are those matrices with a determinant of 1.

Notice that in both Example 18.1.1 and 18.1.2 the identity element of $G$ maps into the identity element of $H$. This is a general result, proved in Theorem 61.

**Theorem 61**: If $G$ and $H$ are two groups and $f$ is a homomorphism, then $f(e_G) = e_H$, and if $f(g) = h$, then $f(g^{-1}) = h^{-1}$.

**Proof**: In Theorem 35 about isomorphisms, the same result was proved without using the fact that an isomorphism is a bijection. Therefore the same proof is valid here. ∎

## ■ *Example 18.1.3*

Let $G$ and $H$ be any groups. Then the function $f : G \to H$ defined by $f(x) = e_H$ for every $x \in G$ is a homomorphism because $f(x)f(y) = e_H e_H = e_H = f(xy)$. This is called the **trivial** homomorphism.

*There may not be a non-trivial homomorphism between two groups, as Example 18.1.4 shows.*

## ■ *Example 18.1.4*

Suppose that you are looking for a homomorphism from $\mathbf{Z}_3$ to $\mathbf{Z}_2$. The identity element of $\mathbf{Z}_3$ must map to the identity element of $\mathbf{Z}_2$, but what is the image of $1 \in \mathbf{Z}_3$? Suppose that $f(1) = 0$. Then $f(2) = f(1+1) = f(1) + f(1) = 0 + 0 = 0$, so you have the trivial homomorphism. Suppose now that $f(1) = 1$, the only other alternative. Then $f(2) = f(1+1) = f(1) + f(1) = 1 + 1 = 0$. But then $f(0) = f(2+1) = f(2) + f(1) = 0 + 1 = 1$ which is a contradiction. Therefore the trivial homomorphism is the only homomorphism between $\mathbf{Z}_3$ to $\mathbf{Z}_2$.

## ■ *Example 18.1.5*

Consider the function $f : S_n \to (\{1, -1\}, \times)$ defined by $f(x) = 1$ if $x$ is an even permutation, and $f(x) = -1$ if $x$ is an odd permutation. It is easy to show that this function is a homomorphism. Notice that it is the even permutations which map on to the identity element.

## ■ *Example 18.1.6*

Consider the function $f : \mathbf{Z}_4 \to \mathbf{Z}_2$ defined by the equation $f([n]_4) = [n]_2$. To prove that this is a homomorphism, notice that:

$$f\big([m]_4 + [n]_4\big) = f\big([m+n]_4\big)$$
$$= [m+n]_2$$
$$= [m]_2 + [n]_2$$
$$= f\big([m]_4\big) + f\big([n]_4\big)$$

## ■ *Example 18.1.7*

Let $f_1 : G_1 \to H_1$ and $f_2 : G_2 \to H_2$ be homomorphisms of groups. Consider the function $f : G_1 \times G_2 \to H_1 \times H_2$ defined by $f(g_1, g_2) = \big(f_1(g_1), f_2(g_2)\big)$. This is a homomorphism because:

$$f\big((x_1, x_2)(y_1, y_2)\big) = f(x_1 y_1, x_2 y_2) \qquad \text{Combining elements in } G_1 \times G_2$$
$$= \big(f_1(x_1 y_1), f_2(x_2 y_2)\big) \qquad \text{Definition of } f$$
$$= \big(f_1(x_1) f_1(y_1), f_2(x_2) f_2(y_2)\big) \qquad f_1, f_2 \text{ h'morphisms}$$
$$= \big(f_1(x_1), f_2(x_2)\big)\big(f_1(y_1), f_2(y_2)\big) \qquad \text{In } H_1 \times H_2$$
$$= f(x_1, x_2) f(y_1, y_2) \qquad \text{Definition of } f$$

In particular, this result, together with Example 18.1.6, shows that $f : \mathbf{Z}_4 \times \mathbf{Z}_4 \to \mathbf{Z}_2 \times \mathbf{Z}_2$ given by $f\big([m]_4, [n]_4\big) = \big([m]_2, [n]_2\big)$ is a homomorphism. The elements which map to the identity in $\mathbf{Z}_2 \times \mathbf{Z}_2$ satisfy $f\big([m]_4, [n]_4\big) = \big([0]_2, [0]_2\big)$, that is, $\big([m]_2, [n]_2\big) = \big([0]_2, [0]_2\big)$. These elements form the set $\{(0,0), (2,0), (0,2), (2,2)\}$ in $\mathbf{Z}_4 \times \mathbf{Z}_4$.

## ■ *Example 18.1.8*

Let $N$ be a normal subgroup of a group $G$. Then the function $f : G \to G/N$ defined by $f(x) = xN$ for $x \in G$ is a homomorphism.

Let $x$ and $y$ be any members of $G$. Then $f(x)f(y) = (xN)(yN)$ by definition of $f$, and $(xN)(yN) = (xy)N$ by definition of the group operation in $G/N$. But $(xy)N = f(xy)$ by definition of $f$. Therefore $f(x)f(y) = f(xy)$ as required.

The identity element of $G/N$ is the coset $N$. The elements which map on to $N$ are the solutions of the equation $f(x) = N$, that is the elements $x$ such that $xN = N$. By Theorem 46, part (1) $xN = yN$ if, and only if, $x^{-1}y \in N$. Putting $x = e$, $yN = N$ if, and only if $y \in N$. The solution set is therefore $N$.

*The importance of this example is the subject of a comment in the next section.*

## 18.2 THE KERNEL OF A HOMOMORPHISM

In a number of the Examples 18.1.1 to 18.1.7, you have seen that the elements which are mapped to the identity element by a homomorphism form a subgroup of the domain.

In Example 18.1.1, the subgroup was $4\mathbf{Z}$, while in Example 18.1.2 it was the subgroup of all matrices with determinant 1.

**Definition**: The **kernel of a homomorphism** $f$ of a group $G$ to a group $H$ is the set of all elements of $G$ which are mapped by $f$ into the identity element of $H$. The kernel of $f$ is written $\ker f$.

Here is a proof that the kernel of a homomorphism is a subgroup of the domain, in fact a normal subgroup.

**Theorem 62**: Let $f : G \to H$ be a homomorphism from a group $G$ into a group $H$. Then the kernel of $f$ is a normal subgroup of $G$.

*In the proof, the conditions of Theorem 21 are first used to show that the kernel is a subgroup.*

**Proof**: From Theorem 61, $f(e_G) = e_H$, so $e_G \in \ker f$.

Suppose that $x, y \in \ker f$. Then $f(x) = f(y) = e_H$. It follows that $f(xy) = f(x)f(y) = e_H e_H = e_H$ showing that $xy \in \ker f$.

Similarly, from Theorem 61, $f(x^{-1}) = f(x)^{-1} = e_H^{-1} = e_H$, showing that $f(x^{-1}) = e_H$ and $x^{-1} \in \ker f$.

It follows that $\ker f$ is a subgroup of $G$.

Suppose now that $x \in G$ and $k \in \ker f$. To prove that $x^{-1}kx \in \ker f$, you need to show that it maps under $f$ to the identity in $H$. So $f(x^{-1}kx) = f(x^{-1})f(k)f(x) = f(x^{-1})e_H f(x)$ and $f(x^{-1})e_H f(x) = f(x^{-1})f(x) = f(x^{-1}x) = f(e_G) = e_H$. It follows that $x^{-1}kx \in \ker f$. So $\ker f$ is a normal subgroup of $G$. ∎

*It should not surprise you by now that the image of a subgroup is a subgroup. This is given as an exercise, and will now be assumed.*

*While Theorem 62 says that every kernel is a normal subgroup, the importance of Example 18.1.8 is that it shows that the reverse is true, that is, every normal subgroup N of a group G is the kernel of some homomorphism; it is the kernel of the homomorphism $f : G \to G/N$ defined by $f(x) = xN$ for $x \in G$.*

Suppose now that two elements $x \in G$ and $y \in G$ map to the same image under a homomorphism $f : G \to H$. What can you say about $x$ and $y$? The answer is that they belong to the same coset of ker $f$ in $G$.

**Theorem 63**: Let $f : G \to H$ be a homomorphism of a group $G$ into a group $H$. Then $f(x) = f(y)$ if, and only if, $x$ and $y$ belong to the same coset of ker $f$ in $G$.

*If.* Let $K = \ker f$. Suppose that $x$ and $y$ are in the same coset of $K$. Then $xK = yK$, so, by Theorem 46 part (1), $x^{-1}y \in K$. Then $f(x)^{-1}f(y) = f(x^{-1})f(y) = f(x^{-1}y) = e_H$, and so $f(x) = f(y)$.

*Only if.* Let $K = \ker f$. Suppose that $f(x) = f(y)$. Then $f(x^{-1}y) = f(x^{-1})f(y) = f(x)^{-1}f(y) = e_H$. This shows that $x^{-1}y \in K$ and therefore that $xK = yK$. ∎

## WHAT YOU SHOULD KNOW

- What a homomorphism between groups is.

- That the identity maps to the identity, and inverses map to inverses.

- What the kernel of a homomorphism is.

- That the kernel of a homomorphism is a normal subgroup.

- That two elements map to the same element under a homomorphism if, and only if, they belong to the same coset of the kernel.

## EXERCISE 18

**1** Let $f : G \to H$ be a homomorphism from a group $G$ to a group $H$. Prove that the image of a subgroup of $G$ is a subgroup of $H$.

**2** Let $G$ be a cyclic group, and let $f : G \to H$ be a surjective homomorphism. Prove that $H$ is a cyclic group.

**3** Let $G$ be a group and let $a$ be any element of $G$. Prove that $f : (\mathbf{Z}, +) \to G$ defined by $f(n) = a^n$ is a homomorphism.

**4** Determine which for the following functions from groups $G \to H$ are homomorphisms. For those which are homomorphisms, determine the kernel.

   (1)   $f : (\mathbf{R}^*, \times) \to (\mathbf{R}^*, \times)$, where $f(x) = |x|$

   (2)   $f : (\mathbf{R}, +) \to (\mathbf{Z}, +)$, where $f(x) =$ the largest integer $\leq x$

   (3)   $f : (\mathbf{R}, +) \to (\mathbf{Z}, +)$, where $f(x) = x + 1$

   (4)   $f : (\mathbf{R}^*, \times) \to (\mathbf{R}^*, \times)$, where $f(x) = 1/x$

**5** Let $f : G \to H$ and $g : H \to K$ be homomorphisms of groups. Prove that $gf : G \to K$ is a homomorphism.

**6** Let $f : G \to H$ be a homomorphism of groups. Prove that $f$ is an injection if, and only if, $\ker f = \{e_G\}$.

# 19

## *The first isomorphism theorem*

### 19.1 MORE ABOUT THE KERNEL

Suppose that $f : G \to H$ is a homomorphism of groups and also a surjection.

In Chapter 17, you saw that the kernel of a homomorphism consisted of those elements which map onto the identity element, and that the solution for $x \in G$ of an equation such as $f(x) = a$ is a coset in $G$. You find one element $g \in G$ such that $f(g) = a$, and then you find the coset $aK$, where $K = \ker f$.

It is quite likely that you have seen this situation before, but not in abstract form.

### ■ *Example 19.1.1*

Suppose that you wish to solve the complex equation $z^3 = 8i$. One method is first to find a solution of $z^3 = 8i$, and then multiply it by all the solutions of the equation $z^3 = 1$.

This is a concrete example of a group homomorphism. In $(\mathbf{C}^*, \times)$, consider the function $f : (\mathbf{C}^*, \times) \to (\mathbf{C}^*, \times)$ given by $f(z) = z^3$. It is a homomorphism because $f(z_1)f(z_2) = z_1^3 z_2^3 = (z_1 z_2)^3 = f(z_1 z_2)$ for any complex numbers $z_1$ and $z_2$.

In this case, one root of $z^3 = 8i$ is $2\left(\cos\frac{1}{6}\pi + i\sin\frac{1}{6}\pi\right)$. The kernel $K$ of the homomorphism are the solutions of the equation $z^3 = 1$, that is, $K = \left\{1, \cos\frac{2}{3}\pi + i\sin\frac{2}{3}\pi, \cos\frac{4}{3}\pi + i\sin\frac{4}{3}\pi\right\}$. So the solutions of $z^3 = 8i$ are the elements of the coset $2\left(\cos\frac{1}{6}\pi + i\sin\frac{1}{6}\pi\right)K$, that is $\left\{2\left(\cos\frac{1}{6}\pi + i\sin\frac{1}{6}\pi\right), 2\left(\cos\frac{5}{6}\pi + i\sin\frac{7}{6}\pi\right), 2\left(\cos\frac{3}{2}\pi + i\sin\frac{3}{2}\pi\right)\right\}$. The third root is actually $-2i$.

> Note that you would have got the same result if you had noticed that one root of $z^3 = 8i$ is $z = -2i$, and then multiplied it by all the members of the kernel to get the coset containing $-2i$.

## ■ *Example 19.1.2*

Suppose that $(\mathbf{D}, +)$ is the set of twice differentiable functions. This is a group under addition of functions defined by the equation $(f_1 + f_2)(x) = f_1(x) + f_2(x)$. Consider the differentiation function $d : (\mathbf{D}, +) \to (\mathbf{D}, +)$ defined by $d(f) = df/dx$.

Notice first that the function $d : (\mathbf{D}, +) \to (\mathbf{D}, +)$ is a homomorphism, because $d(f)_1 + d(f_2) = df_1/dx + df_2/dx = d(f_1 + f_2)/dx = d(f_1 + f_2)$.

Now consider the equation $df/dx = 2x$. First you find any solution of the equation $df/dx = 2x$, say $x^2$. Then you find the kernel of $d$, that is, the solutions of the equation $df/dx = 0$. The kernel is given by $K = \mathbf{R}$, since constants, and only constants, differentiate to give 0. Combining the solution $f = x^2$ with the kernel, you find that all solutions are given by $x^2 + c$, where $c$ is a real constant.

## ■ *Example 19.1.3*

This is really an extension of Example 19.1.2. Suppose that $(\mathbf{D}, +)$ is the group of twice-differentiable functions under addition. Consider the function $d : (\mathbf{D}, +) \to (\mathbf{D}, +)$ defined by $d(f) = df/dx + f$.

Now consider the equation $df/dx + f = 2$. First you find any solution of the equation $df/dx + f = 2$, say $f(x) = 2$. Then you find the kernel of $d$, that is, the solutions of the equation $df/dx + f = 0$, that is,

$f(x) = Ae^{-x}$, where $A \in \mathbf{R}$. Combining the solution $f(x) = 2$ with the kernel, you find that all solutions are given by $f(x) = Ae^{-x} + 2$.

> When solving differential equations, the kernel is often called the complementary function, and the other solution is called the particular integral.
>
> In Examples 19.1.2 and 19.1.3, it is not suggested that you solve these particular equations by the method involving cosets. However, it is always useful to be able to see elementary and familiar methods in a wider, abstract, context. It helps to understand the abstract context better. In Examples 19.1.2 and 19.1.3, to be sure that you have all the solutions, you do actually use the group theory theorem.

## 19.2 THE QUOTIENT GROUP OF THE KERNEL

The kernel of a homomorphism of groups is a normal subgroup. (See Theorem 62.) Therefore the set of cosets of the kernel forms a quotient group under the operation induced by the group operation.

### ■ *Example 19.2.1*

In Example 18.1.1, you considered the function from $(\mathbf{Z}, +)$ to the group $(\{1, i, -1, -i\}, \times)$ defined by $f(n) = i^n$. You also showed that this function was a homomorphism. The kernel $K = \{n \in \mathbf{Z} : i^n = 1\}$, which clearly consists of the multiples of 4, is $4\mathbf{Z}$. The quotient group is $\mathbf{Z}/4\mathbf{Z}$, which has the four elements, $4\mathbf{Z}$, $1 + 4\mathbf{Z}$, $2 + 4\mathbf{Z}$ and $3 + 4\mathbf{Z}$.

### ■ *Example 19.2.2*

In Example 18.1.7, you saw that the function $f : \mathbf{Z}_4 \times \mathbf{Z}_4 \to \mathbf{Z}_2 \times \mathbf{Z}_2$ given by $f([m]_4, [n]_4) = ([m]_2, [n]_2)$, was a homomorphism, and that the kernel $K = \{([0]_4, [0]_4), ([2]_4, [0]_4), ([0]_4, [2]_4), ([2]_4, [2]_4)\}$. There are 16 elements in $\mathbf{Z}_4 \times \mathbf{Z}_4$, and four in $K$ so the number of elements in the quotient group $(\mathbf{Z}_4 \times \mathbf{Z}_4)/K$ is $16/4 = 4$. Four cosets are $K$, $([1]_4, [0]_4) + K$, $([0]_4, [1]_4) + K$ and $([1]_4, [1]_4) + K$. These cosets are distinct, for suppose that $([i]_4, [j]_4) + K = ([k]_4, [l]_4) + K$ where

$i, j, k, l \in \{0,1\}$. Then, by Theorem 46, $\left(\left[k-i\right]_4, \left[l-j\right]_4\right)$ $= \left(\left[k\right]_4, \left[l\right]_4\right) - \left(\left[i\right]_4, \left[j\right]_4\right) \in K$. But satisfying both $\left(\left[k-i\right]_4, \left[l-j\right]_4\right) \in K$ and $i, j, k, l \in \{0,1\}$ is impossible unless $i = k$ and $j = l$. Therefore the four cosets are distinct, and are the elements of the quotient group $\left(\mathbf{Z}_4 \times \mathbf{Z}_4\right)/K$.

This is not a cyclic group since every element has order 2. So it is isomorphic to the only other group of order 4, namely $\mathbf{Z}_2 \times \mathbf{Z}_2$. Therefore $\left(\mathbf{Z}_4 \times \mathbf{Z}_4\right)/K \cong \mathbf{Z}_2 \times \mathbf{Z}_2$.

## 19.3 THE FIRST ISOMORPHISM THEOREM

You may be able to guess at the theorem which is coming, from Examples 19.2.1 and 19.2.2. If you have a surjective group homomorphism $f$ from a group $G$ to a group $H$, then the group of cosets of the kernel is isomorphic to the image.

You may find Fig. 19.1 helpful. On the left you see four elements of $G$, the identity $e$, two elements $x$ and $y$, and their product $xy$. On the right are the images in $H$ of these four elements under the homomorphism. The homomorphism relation $f(xy) = f(x)f(y)$ becomes, in this case, $f(xy) = st$, exhibiting the fact that the product $xy$ in $G$ maps to the product $st$ of $s$ and $t$ in $H$.

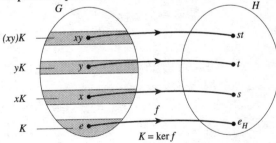

**Fig. 19.1**

The shaded region at the bottom of $G$ represents $\ker f$, written as $K$. The shaded piece $xK$ is the coset of $K$ in $G$ which contains $x$. Similarly shaded are the cosets containing $y$ and $xy$, namely $yK$ and $(xy)K$. Notice that these cosets, which are in the quotient group $G/K$, combine according to the rule $(xK)(yK) = (xy)K$.

So the group of shaded cosets on the left of the diagram is isomorphic to the image on the right of the diagram.

Here is a formal proof of the theorem, which is sometimes called the first isomorphism theorem.

> *There is some variation in the terminology here. There are several isomorphism theorems, (only one of which will be in this book), and some authors give the name First Isomorphism Theorem to one of the others.*

**Theorem 64**: **The first isomorphism theorem**: Let $f : G \to H$ be a homomorphism of groups. Then $G/\ker f \cong \operatorname{im} f$.

**Proof**: Let $K = \ker f$ and the image of $f$ be $\operatorname{im} f$. First note that as $K$ is a normal subgroup, by Theorem 62, $G/K$ is a group, and that $\operatorname{im} f$ is a group because it is the image of a group under a homomorphism, Exercise 17, question 1.

> *The proof now comes from pulling together two separate strands: the work on equivalence classes, in particular Theorem 57, and Theorem 63 on homomorphisms.*

Applying the work of Section 16.4 to the function $f : G \to H$, there is an equivalence relation on $G$ given by $x \sim y$ if, and only if, $f(x) = f(y)$, and a bijection $\theta : G/\sim \to \operatorname{im} f$ given by $\theta(\bar{x}) = f(x)$ for $x \in G$. But, by Theorem 63, the equivalence classes of $\sim$ are the cosets of $K = \ker f$; that is, $\bar{x} = xK$ for each $x \in G$. So $\theta$ is the function $G/K \to \operatorname{im} f$ given by $\theta(xK) = f(x)$

You now have to show that $\theta$ is an isomorphism by showing that it has the homomorphism property that $\theta\big((xK)(yK)\big) = \theta(xK)\theta(yK)$.

$$\theta\big((xK)(yK)\big) = \theta\big((xy)K\big) \qquad \text{Combining cosets in } G$$
$$= f(xy) \qquad \text{Using the definition of } \theta$$
$$= f(x)f(y) \qquad f \text{ is a homomorphism}$$
$$= \theta(xK)\theta(yK) \qquad \text{The definition of } \theta$$

Therefore $\theta$ is an isomorphism. ∎

*This theorem throws up pairs of groups that are isomorphic. Some of the pairs may be surprising.*

## ■ Example 19.3.1

Consider $(\mathbf{Z},+)$ and $(\mathbf{Z}_n,+)$ and the function $f:(\mathbf{Z},+) \to (\mathbf{Z}_n,+)$ defined by $f(m)=[m]$. This function is a homomorphism since $f(m_1)+f(m_2)=[m_1]+[m_2]=[m_1+m_2]=f(m_1+m_2)$. The image of $f$ is the whole of $\mathbf{Z}_n$. The kernel is the subgroup $n\mathbf{Z}$. So, by the first isomorphism theorem, $\mathbf{Z}/n\mathbf{Z} \cong \mathbf{Z}_n$.

## ■ Example 19.3.2

Let $(\mathbf{M},\times)$ be the group of invertible $2 \times 2$ matrices with real entries under matrix multiplication and $(\mathbf{R}^*,\times)$ be the set of non-zero real numbers under multiplication.

In Example 18.1.2, you showed that $f:(\mathbf{M},\times) \to (\mathbf{R}^*,\times)$ defined by $f(A) = \det A$ was a homomorphism. The kernel of $f$ is the set $\mathbf{U}$ of matrices with determinant 1. The image is $\mathbf{R}^*$, as, given a real number $z$, $\det\begin{pmatrix} z & 0 \\ 0 & 1 \end{pmatrix} = z$. By the first isomorphism theorem, $\mathbf{M}/\mathbf{U} \cong \mathbf{R}^*$.

## ■ Example 19.3.3

Let $f:(\mathbf{R},+) \to (\mathbf{C}^*,\times)$ be defined by $f(x)=e^{2\pi i x}$. This is a homomorphism because $f(x+y)=e^{2\pi i(x+y)}=e^{2\pi i x+2\pi i y}=f(x)f(y)$.

The kernel is the set of real numbers $x$ such that $e^{2\pi i x}=1$, that is $\mathbf{Z}$. The image is the set $\mathbf{T}$ of complex numbers with unit modulus, that is, the set of complex numbers which lie on the unit circle. So $\mathbf{R}/\mathbf{Z} \cong \mathbf{T}$, the set of complex numbers on the unit circle. The first isomorphism theorem indicates that the operation of complex number multiplication on numbers of unit modulus is the same as adding the angles at the centre of the circle. You can also think of this as the real axis being bent round the unit circle.

## ■ *Example 19.3.4*

Let $G$ be a group, and let $a$ be any element of $G$. Then, from question 3, Exercise 18, $f:(\mathbf{Z},+) \to G$ given by $f(n) = a^n$ is a group homomorphism. The image of $f$ is $\langle a \rangle$, and $\ker f = \{m \in \mathbf{Z} : a^m = e\}$.

If $a$ has infinite order $\ker f = \{0\}$, by Theorem 18, part (1).

If $a$ has finite order $n$, then $\ker f = n\mathbf{Z}$, by Theorem 18, part (2).

Hence, if $G$ is cyclic, (so that $f$ is surjective), then, by the first isomorphism theorem:

> if $G$ is infinite cyclic, then $G \cong \mathbf{Z}$
>
> if $G$ is finite cyclic, order $n$, then $G \cong \mathbf{Z}/n\mathbf{Z}$

This is an alternative way of looking at the result of Theorem 36.

## WHAT YOU SHOULD KNOW

- How to solve equations involving homomorphisms.

- The first isomorphism theorem and its meaning.

- How to prove that two groups are isomorphic, when one of them is a quotient group.

## EXERCISE 19

**1** Let $S_n$ be the group of permutations on $n$ symbols, and let $A_n$ be the alternating subgroup of $S_n$. Show that $A_n$ is a normal subgroup of $S_n$, and that $S_n/A_n \cong (\{1, -1\}, \times)$.

**2** Find a surjective homomorphism $f:(\mathbf{C}^*, \times) \to (\mathbf{R}^+, \times)$, and then apply the first isomorphism theorem to it.

**3** Let **a** be a fixed non-zero vector in three-dimensional space $\mathbf{R}^3$, and let $f:(\mathbf{R}^3, +) \to (\mathbf{R}, +)$ be defined by $f(\mathbf{x}) = \mathbf{a.x}$, the scalar product of **x** with **a**. Show that $f$ is a homomorphism, and identify the conclusion of the first isomorphism theorem.

**4** Let $G$ be the group $(\{a + bi : a, b \in \mathbf{Z}\}, +)$ and let $H$ be the normal subgroup $(\{2a + 2bi : a, b \in \mathbf{Z}\}, +)$. Identify $G/H$.

**5** Let $p$ and $q$ be positive integers. Prove that $\mathbf{Z}_{pq}/(p\mathbf{Z})_{pq} \cong \mathbf{Z}_p$, where $(p\mathbf{Z})_{pq} = \{[0]_{pq}, [p]_{pq}, [2p]_{pq}, \ldots, [(q-1)p]_{pq}\}$.

# 20

# *Answers*

## CHAPTER 1

**1** Let the odd numbers be $2m+1$ and $2n+1$. Then the product is $(2m+1)(2n+1) = 4mn+2m+2n+1 = 2(2mn+m+n)+1$, which is an odd number.

**2** Suppose that there are positive whole numbers $m$ and $n$, with $m > n$, such that $m^2 - n^2 = 6$. Then $(m-n)(m+n) = 6$. The only factors of 6 are 1 and 6, or 2 and 3, so either $(m+n) = 6$ and $(m-n) = 1$ or $(m+n) = 3$ and $(m-n) = 2$. Thus either $m = 3\frac{1}{2}$, $n = 2\frac{1}{2}$, or $m = 2\frac{1}{2}$, $n = \frac{1}{2}$. Either way leads to a contradiction, so the supposition was false.

**3** $n^2 + n = n(n+1)$. This is the product of two consecutive numbers, so one of them is even. The product is therefore even.

**4** You would need to turn over the first to check that it was not striped, and the last to check that it is a diamond. (The second card doesn't matter, because it could be either chequered or striped: the third doesn't matter because, as it is chequered, it could be a diamond or a spade.) If the first card is not striped and the last is a diamond, then the statement is true. If either the first card is striped or the last is not a diamond, then the statement is false.

**5** If you attempt to follow through with $\sqrt{4} = a/b$, you get $a^2 = 4b^2$, you can deduce that $a^2$ is a multiple of 4, but you cannot go on to deduce that $a$ is a multiple for 4. It breaks down because 4 is not prime; see Theorem 4. A simple counterexample is 4 divides $6^2$, but 4 does not divide 6.

**6**     Consider $x = -2$. Then $-2 < 1$ is true, but $(-2)^2 = 4$ and $4 > 1$. So $x = -2$ is a counterexample.

**7**     If $\sqrt{a+b} = \sqrt{a} + \sqrt{b}$, then, by squaring, $a + b = a + 2\sqrt{ab} + b$ or $\sqrt{ab} = 0$. Therefore $ab = 0$, so either $a = 0$ or $b = 0$.

**8**     *If.* Suppose that $pq$ is odd, and that one of $p$ and $q$, say $p$ is even. Then $p = 2m$ for some integer $m$, and $pq = 2mq = 2(mq)$ is divisible by 2, which is a contradiction. Therefore $p$ and $q$ are both odd. *Only if.* See question 1.

**9**     Let $N = a_n 10^n + a_{n-1} 10^{n-1} + \ldots + a_1 10 + a_0$ where $0 \le a_i < 10$ for all the coefficients $a_i$. *Necessary.* Suppose that 9 divides $N$. Then 9 divides $a_n 10^n + \ldots + a_1 10 + a_0$. But for all $i$, $10^i$ leaves remainder 1 on division by 9. So the remainder when $N$ is divided by 9 is $a_n + \ldots + a_0$. But as 9 divides $N$ the remainder is 0, so $a_n + \ldots + a_0$ is divisible by 9, and the sum of the digits is divisible by 9. *Sufficient.* Suppose that 9 divides the sum of the digits, that is 9 divides $a_n + \ldots + a_0$. Then 9 also divides the sum

$$\left(a_n + \ldots + a_0\right) + \left(\overbrace{99\ldots9}^{n\ 9s} a_n + \ldots + 9a_1\right) = N.$$

## CHAPTER 2

**1**     (1) Properly defined. (2) This is debatable. April 1st means a different time in Japan from the time in the USA. If these differences are resolved, and other issues(!), such as whether the time of birth is exact, can be settled, the set may be well defined. (3) This is well defined, although you might have difficulty in finding $A$.

**2**     (1) T. (2) F. See definition. (3) T.

**3**     $\{-2, , 1, 0, 1, 2\}$ No.

**4**     Let $x \in \mathbf{Z}$. Then $x = x/1$ where $x \in \mathbf{Z}$ and $1 \in \mathbf{Z}^*$, so $x \in \mathbf{Q}$.

**5**     $2\mathbf{Z} \subseteq \mathbf{Z}$ is true, because if $y \in 2\mathbf{Z}$ then $y = 2x$ for some $x \in \mathbf{Z}$, so $y \in \mathbf{Z}$. $\mathbf{Z} \subseteq 2\mathbf{Z}$ is not true; for example, $3 \in \mathbf{Z}$, but $3 \notin 2\mathbf{Z}$. $2\mathbf{Z}$ is the set of even numbers.

**6**     *Part 1:* Proof that if $A \cup B = B$ then $A \cap B = A$. Suppose that $x \in A \cap B$. Then $x \in A$ (by definition). Hence $A \cap B \subseteq A$. Now suppose that $x \in A$. Then $x \in A \cup B$ (by definition), so $x \in B$ (by hypothesis). Therefore $x \in A$ and $x \in B$, so $x \in A \cap B$. Hence $A \subseteq A \cap B$. Therefore $A \cap B = A$.

*Part* 2: Proof that if $A \cap B = A$ then $A \cup B = B$. Suppose that $x \in A \cup B$. Then $x \in A$ or $x \in B$ (by definition). Suppose that $x \in A$. Then, since $A \cap B = A$, $x \in A \cap B$ so $x \in B$. Either way $x \in B$, so $A \cup B \subseteq B$. Now suppose that $x \in B$. Then, $x \in A \cup B$ (by definition). So $B \subseteq A \cup B$. Therefore $A \cup B = B$. Therefore the statements $A \cup B = B$ and $A \cap B = A$ are equivalent.

*In fact, these two statements are also equivalent to the statement $A \subseteq B$. See Example 2.7.1.*

**7** *Part* 1: Proof that $A \cap (B \cup C) \subseteq (A \cap B) \cup (A \cap C)$. Suppose that $x \in A \cap (B \cup C)$. Then $x \in A$ and $x \in B \cup C$. Therefore $x \in A$ and ($x \in B$ or $x \in C$). So either ($x \in A$ and $x \in B$) or ($x \in A$ and $x \in C$). Hence $x \in (A \cap B) \cup (A \cap C)$, so $A \cap (B \cup C) \subseteq (A \cap B) \cup (A \cap C)$.

*Part* 2: Proof that $(A \cap B) \cup (A \cap C) \subseteq A \cap (B \cup C)$. Suppose that $x \in (A \cap B) \cup (A \cap C)$. Then either ($x \in A$ and $x \in B$) or ($x \in A$ and $x \in C$), so $x \in A$ and ($x \in B$ or $x \in C$). Therefore $x \in A \cap (B \cup C)$.

**8**  (1) T. (2) F. In fact, if $A \subseteq (A \cap B)$, then $A \subseteq B$. (3) T. (4) T. (5) F. This statement is true for some sets $A$. For instance, suppose $A = \{\text{abstract ideas}\}$; then $A \in A$. (6) F. See definition. (7) F. In fact, $(A \cup B) \subseteq (A \cap B)$ if, and only if, $A = B$.

*Part 5 is close to Russell's paradox. Let $X = \{A : A \notin A\}$. Then $X \in X$ if, and only if, $X \notin X$. The barber paradox is: The barber shaves everyone who does not shave himself. The barber does not shave anyone who does shave himself. Who shaves the barber? The answer is that the situation cannot exist. Also, the situation of the 'set of all sets' cannot exist (without modification).*

**9**  Suppose that $x \in B$. Then $|x - 1| < 2$, so $-2 < x - 1 < 2$ or $-1 < x < 3$. Hence $-3 < x < 3$ (this is a weaker inequality.) So $x \in A$. Therefore $B \subseteq A$.

**10**  The statement is true when $A$ has one element $a$. There are two subsets, $\varnothing$ and $A$. Suppose that the statement is true for $n = k$, and consider the set consisting of $k + 1$ elements, $a_1, a_2, \ldots, a_k$, together with the element $x$. Then to every one of the $2^k$ subsets of $\{a_1, a_2, \ldots, a_k\}$, you can either add $x$, or not. There are thus $2 \times 2^k = 2^{k+1}$ subsets of a set with $k + 1$ elements. So, if the statement is true for $k$ elements, it is also true for $k + 1$. Therefore, by the principle of mathematical induction, it is true for all positive integers.

# CHAPTER 3

**1**  (1) Binary operation. (2) Binary operation. (3) Not a binary operation; you cannot divide by zero. (4) Not a binary operation;

$2 \lozenge (-1)$ is not defined in $\mathbf{Z}$. (5) Binary operation. (6) Binary operation. (7) Binary operation. (8) Binary operation. (9) Not a binary operation; not all matrices are conformable for multiplication. (10) Not a binary operation; $\det(A - B)$ is not a matrix. (11) Not a binary operation; $(-1) \lozenge \frac{1}{2}$ is not defined.

## CHAPTER 4

**1**   Suppose that $m = 6$, $n = 4$ and $a = 3$. Then $m$ and $n$ are not relatively prime. Since 6 divides 12, $m$ divides $na$, but 6 does not divide 3, so $m$ does not divide $a$. Therefore Theorem 2 is not true if $m$ and $n$ are not relatively prime.

**2**   Suppose that $m = 6$, $n = 10$ and $k = 30$. Then $m$ divides $k$ and $n$ divides $k$. However, $mn = 60$ does not divide $k = 30$. Therefore Theorem 3 is not true if $m$ and $n$ are not relatively prime.

**3**   (1) T. (2) T. (3) F. ($5 \equiv -13 \pmod 3$), as 3 divides 18. Therefore $5 \not\equiv -13 \pmod 3$ is false.

**4**   (1) 1. (2) 0. (3) 3. (4) 0.

**5**   Testing the numbers 0, 1, 2, ... , 10 in turn, you find that 5 and 6 are the only solutions. Or you could factorise the equation $x^2 - 3 \equiv 0 \pmod{11}$ which is the same as $x^2 + 8 \equiv 0 \pmod{11}$ to get $(x + 6)(x + 5) \equiv 0 \pmod{11}$, so, using Theorem 9, $x \equiv -6 \pmod{11}$ or $x \equiv -5 \pmod{11}$ leading to $x \equiv 5$ or $x \equiv 6$ as before.

*Notice that Theorem 9 guarantees that there are only two solutions, because the modulus, 11, is prime.*

**6**   Notice first that you can have the same set of $n$ primes for both $a$ and $b$, by allowing some of the $r$s and $s$s to be zero. Thus, $589 = 19 \times 31$ and $380 = 2^2 \times 5 \times 19$; however, $589 = 2^0 \times 5^0 \times 19 \times 31$ and $380 = 2^2 \times 5 \times 19 \times 31^0$, so they have the same set of $n = 4$ primes. Now, using the hint, $h = p_1^{\alpha_1} \ldots p_n^{\alpha_n}$ and $l = p_1^{\beta_1} \ldots p_n^{\beta_n}$ where $\alpha_i = \min\{r_i, s_i\}$ and $\beta_i = \max\{r_i, s_i\}$, $i = 1, \ldots, n$. Therefore $hl = p_1^{\alpha_1} \ldots p_n^{\alpha_n} p_1^{\beta_1} \ldots p_n^{\beta_n} = p_1^{\alpha_1 + \beta_1} \ldots p_n^{\alpha_n + \beta_n}$.   However, $ab = p_1^{r_1} \ldots p_n^{r_n} p_1^{s_1} \ldots p_n^{s_n} = p_1^{r_1 + s_1} \ldots p_n^{r_n + s_n}$. Looking at the indices, you need to prove that $\alpha_i + \beta_i = r_i + s_i$, for $i = 1, \ldots, n$, that is, $\min\{r_i, s_i\} + \max\{r_i, s_i\} = r_i + s_i$ for $i = 1, \ldots, n$. Suppose that $r_i \geq s_i$. Then   $\min\{r_i, s_i\} = s_i$   and   $\max\{r_i, s_i\} = r_i$,   so   that $\min\{r_i, s_i\} + \max\{r_i, s_i\} = r_i + s_i$. If, on the other hand, $r_i < s_i$, then $\min\{r_i, s_i\} = r_i$ and $\max\{r_i, s_i\} = s_i$, once again giving the result $\min\{r_i, s_i\} + \max\{r_i, s_i\} = r_i + s_i$. As this is true, $ab = hl$.

## CHAPTER 5

**1**

|     | 1  | 5  | 7  | 11 |
|-----|----|----|----|----|
| 1   | 1  | 5  | 7  | 11 |
| 5   | 5  | 1  | 11 | 7  |
| 7   | 7  | 11 | 1  | 5  |
| 11  | 11 | 7  | 5  | 1  |

It is a group table.

**2**

|     | $I$ | $F$ | $G$ | $H$ | $K$ | $L$ |
|-----|-----|-----|-----|-----|-----|-----|
| $I$ | $I$ | $F$ | $G$ | $H$ | $K$ | $L$ |
| $F$ | $F$ | $G$ | $I$ | $L$ | $H$ | $K$ |
| $G$ | $G$ | $I$ | $F$ | $K$ | $L$ | $H$ |
| $H$ | $H$ | $K$ | $L$ | $I$ | $F$ | $G$ |
| $K$ | $K$ | $L$ | $H$ | $G$ | $I$ | $F$ |
| $L$ | $L$ | $H$ | $K$ | $F$ | $G$ | $I$ |

The closure, identity and inverse properties follow from the work done in producing the table.

**3**    To check whether $\circ$ is a binary operation on $\mathbf{R}-\{-1\}$, notice first that $x \circ y \in \mathbf{R}$. But could $x \circ y = -1$? If $x \circ y = -1$, then $x+y+xy=-1$, $x+y+xy+1=0$, $(x+1)(y+1)=0$ so that $x=-1$ or $y=-1$. Since $-1 \notin \mathbf{R}-\{-1\}$, the operation $\circ$ is closed on $\mathbf{R}-\{-1\}$. The operation $\circ$ is associative; $(x \circ y) \circ z = (x+y+xy) \circ z$ $= x+y+xy+z+xz+yz+xyz$ and also $x \circ (y \circ z) = x \circ (y+z+yz)$ $= x+y+z+yz+xy+xz+xyz$. The identity element is 0, since $x \circ 0$ $= 0 \circ x = x$. Finally, the inverse of $x$ is $-x/(1+x)$, which $\in \mathbf{R}$ because $x \neq -1$, since $x \circ -x/(1+x) = x-x/(1+x)-x^2/(1+x)=0$, after some manipulation, and also $-x/(1+x) \circ x = -x/(1+x)+x-x^2/(1+x)=0$. Notice also that $-x/(1+x) \neq -1$, so $-x/(1+x) \in \mathbf{R}-\{-1\}$.

**4**    Align the rectangle with the axes. Let $I$ be the identity, $X$ be a reflection of the plane in the horizontal axis, $Y$ be a reflection of the plane in the vertical axis and $R$ be a rotation of the plane about the centre of the rectangle of 180°. Then the table is:

|     | $I$ | $X$ | $Y$ | $R$ |
|-----|-----|-----|-----|-----|
| $I$ | $I$ | $X$ | $Y$ | $R$ |
| $X$ | $X$ | $I$ | $R$ | $Y$ |
| $Y$ | $Y$ | $R$ | $I$ | $X$ |
| $R$ | $R$ | $Y$ | $X$ | $I$ |

**5**

|     | 1 | 3 | 7 | 9 |
|-----|---|---|---|---|
| 1   | 1 | 3 | 7 | 9 |
| 3   | 3 | 9 | 1 | 7 |
| 7   | 7 | 1 | 9 | 3 |
| 9   | 9 | 7 | 3 | 1 |

**6**

| | 1 | 2 | 4 | 8 |
|---|---|---|---|---|
| 1 | 1 | 2 | 4 | 8 |
| 2 | 2 | 4 | 8 | 1 |
| 4 | 4 | 8 | 1 | 2 |
| 8 | 8 | 1 | 2 | 4 |

If you re-name the elements 1, 2, 4, 8 as 1, 3, 9, 7, and alter the orders of the rows and columns, the tables become identical. There are also other possibilities for re-naming which make the tables look the same.

**7** If $a^2 = a$, then $a^{-1}(a^2) = a^{-1}a = e$, so $(a^{-1}a)a = e$ and $ea = e$ giving $a = e$. Alternatively, you could use Theorem 17, part (4), and cancel $a$ giving $a = e$ immediately.

**8** Let $a$ and $b$ be any two elements of $G$. Then $ab$ is also an element of $G$, so $(ab)^2 = e$. Therefore $abab = e$. Multiplying this on the left by $a^{-1}$ and on the right by $b^{-1}$ gives $a^{-1}ababb^{-1} = a^{-1}eb^{-1}$ or $ba = a^{-1}b^{-1}$. But since $a^2 = e$, $a = a^{-1}$, and since $b^2 = e$, $b = b^{-1}$. Therefore $a^{-1}b^{-1} = ab$ and $ba = ab$. Therefore $G$ is abelian.

**9** Notice that if a set with a given operation is not a group, it is sufficient to give one reason only. Thus, you may have a different, but equally valid, reason for saying that something is not a group.

(1) Not a group since $1 + 1 = 2$, and the operation is not closed.

(2) This is a group. Call the set G. If $a$ and $b$ are any elements of $G$, $a = 2^m 5^n$ and $b = 2^p 5^q$ for some $m, n, p, q \in \mathbf{Z}$. Therefore $ab = 2^m 5^n 2^p 5^q = 2^{m+p} 5^{n+q}$, and since $m + p, n + q \in \mathbf{Z}$, $ab \in G$. The operation is associative on $G$ since multiplication is associative in $\mathbf{R}$. $1 = 2^0 5^0 \in G$ since $0 \in \mathbf{Z}$. But $1g = g1 = g$ for all $g \in \mathbf{R}$, and $G$ is a subset of $\mathbf{R}$. Finally, if $a = 2^m 5^n \in G$, then $2^{-m} 5^{-n} \in G$. Then $a2^{-m}5^{-n} = 2^m 5^n 2^{-m} 5^{-n} = 2^{m-m} 5^{n-n} = 2^0 5^0 = 1$. Similarly, you find $2^{-m}5^{-n}a = 1$. Therefore $2^{-m}5^{-n}$ is the inverse of $a = 2^m 5^n$ in $G$. Therefore the set of all integers of the form $2^m 5^n$ $(m, n \in \mathbf{Z})$, is a group under multiplication.

(3) Not a group. If you look for an identity element $\begin{pmatrix} x & y \\ 0 & 0 \end{pmatrix}$ such that $\begin{pmatrix} a & b \\ 0 & 0 \end{pmatrix}\begin{pmatrix} x & y \\ 0 & 0 \end{pmatrix} = \begin{pmatrix} a & b \\ 0 & 0 \end{pmatrix}$ for every $\begin{pmatrix} a & b \\ 0 & 0 \end{pmatrix}$, you find that, for $a = 0$ and $b = 1$, $\begin{pmatrix} 0 & 0 \\ 0 & 0 \end{pmatrix} = \begin{pmatrix} 0 & 1 \\ 0 & 0 \end{pmatrix}\begin{pmatrix} x & y \\ 0 & 0 \end{pmatrix} = \begin{pmatrix} 0 & 1 \\ 0 & 0 \end{pmatrix}$. There are no values of $x$ and $y$ which satisfy this equation.

(4)   Not a group. $\begin{pmatrix} 1 & 0 \\ 1 & 1 \end{pmatrix}, \begin{pmatrix} \sqrt{2} & 0 \\ 1 & 1 \end{pmatrix}$ are both members of the set, but

$\begin{pmatrix} 1 & 0 \\ 1 & 1 \end{pmatrix}\begin{pmatrix} \sqrt{2} & 0 \\ 1 & 1 \end{pmatrix} = \begin{pmatrix} \sqrt{2} & 0 \\ 1+\sqrt{2} & 1 \end{pmatrix}$. As $1+\sqrt{2} \notin \mathbf{Q}$, there is no closure.

**10**   You can easily check by writing out all the possibilities that the set is closed, and putting the results in a table. Or you could note that

each of the matrices can be written as $\begin{pmatrix} (-1)^m & 0 \\ 0 & (-1)^n \end{pmatrix}$ for $m,n \in \mathbf{Z}$,

and then write out a general argument, rather like the one in question 9, part 2. In general, matrices are associative under matrix multiplication,

so this set is associative. The matrix $\begin{pmatrix} 1 & 0 \\ 0 & 1 \end{pmatrix}$ acts as the identity, and

from the table you see that all the matrices are self-inverse.

**11**   Let $m$ and $n$ be positive integers. Then

$$x^m x^n = \overbrace{xx\ldots x}^{m\text{ times}}\overbrace{xx\ldots x}^{n\text{ times}} = \overbrace{xx\ldots x}^{m+n\text{ times}} = x^{m+n}.$$

$$\left(x^m\right)^n = \overbrace{\overbrace{xx\ldots x}^{m\text{ times}}\overbrace{xx\ldots x}^{m\text{ times}}\ldots\overbrace{xx\ldots x}^{m\text{ times}}}^{n\text{ times}} = \overbrace{xx\ldots x}^{mn\text{ times}} = x^{mn}$$

$$x^m x^{-n} = \overbrace{xx\ldots xx}^{m\text{ times}}\overbrace{x^{-1}x^{-1}\ldots x^{-1}}^{n\text{ times}}.$$ Suppose first that $m > n$. Then,

cancelling $m-n$ of the products $xx^{-1}$, you reach the result

$x^m x^{-n} = \overbrace{xx\ldots x}^{m-n\text{ times}} = x^{m-n}$. Now suppose $m < n$. Then, cancelling

$n-m$ of the products $xx^{-1}$, you reach the result

$x^m x^{-n} = \overbrace{x^{-1}x^{-1}\ldots x^{-1}}^{n-m\text{ times}}$. But if $m < n$, then $x^{m-n} = \overbrace{x^{-1}x^{-1}\ldots x^{-1}}^{n-m\text{ times}}$, by

definition. Therefore $x^m x^{-n} = x^{m-n}$. Finally, suppose that $m = n$.

Then $x^m x^{-n} = \overbrace{xx\ldots xx}^{m\text{ times}}\overbrace{x^{-1}x^{-1}\ldots x^{-1}}^{m\text{ times}} = e$. But $x^{m-n} = x^{m-m} = x^0 = e$,

so, once again, $x^m x^{-n} = x^{m-n}$. Therefore $x^m x^{-n} = x^{m-n}$ when $m$ and $n$ are positive integers. $x^{-m}x^{-n} = x^{-m-n}$ by arguments similar to the first line.

From all these cases we have that $x^s x^t = x^{s+t}$ for all $s,t \in \mathbf{Z}$.

The result $x^{-m} = \left(x^m\right)^{-1} = \left(x^{-1}\right)^m$ comes directly from the definition.

So $\begin{cases} \left(x^m\right)^{-n} = \left(\left(x^m\right)^n\right)^{-1} = \left(x^{mn}\right)^{-1} = x^{m(-n)} \\ \left(x^{-m}\right)^n = \left(\left(x^m\right)^{-1}\right)^n = \left(x^m\right)^{-n} = x^{m(-n)} = x^{(-m)n} \\ \left(x^{-m}\right)^{-n} = \left(\left(x^{-m}\right)^n\right)^{-1} = \left(x^{(-m)n}\right)^{-1} = x^{-(-m)n} = x^{(-m)(-n)} \end{cases}$

Therefore $\left(x^s\right)^t = x^{st}$ for all $s, t \in \mathbf{Z}$.

Also $x^n = \left(x^{-n}\right)^{-1} = \left(x^{-1}\right)^{-n}$; so $x^{-s} = \left(x^s\right)^{-1} = \left(x^{-1}\right)^s$ for all $s \in \mathbf{Z}$.

**12**  (1) T. See definition; (2) T. For example, the identity. (3) F. $(\mathbf{Z}, +)$ is infinite. (4) F. For example, the group $\{e\}$. (5) F. $e$ has order 1. (6) F. It should be $n$ divides $N$; see Theorem 17. (7) F. See $D_3$ in Example 5.2.2.

**13**  First notice that $\left(x^s\right)^{n/h} = x^{ns/h} = \left(x^n\right)^{s/h} = e^{s/h} = e$. By Theorem 17, part (2), the order of $x^s$ divides $n/h$. Suppose that $m$ is a positive integer such that $\left(x^s\right)^m = e$. Then $x^{sm} = e$, and so $n$ divides $sm$, by Theorem 17, part (2). Therefore $n/h$ divides $(s/h)m$. But $n/h$ and $s/h$ are relatively prime. For if $d > 1$ were a common factor of $n/h$ and $s/h$, then $dh > h$ would be a common factor of $n$ and $s$, contrary to the fact that $h$ is the highest common factor of $n$ and $s$. Therefore, by Theorem 2, $n/h$ divides $m$. So $n/h \leq m$. Therefore $\left(x^s\right)^{n/h}$ is the smallest power of $x^s$ to equal $e$. Therefore the order of $x^s$ is $n/h$.

# CHAPTER 6

**1**  The subgroups are $\{6\}$, $\{6, 4\}$ and $\{6, 4, 2, 8\}$. The only proper subgroup is $\{6, 4\}$.

**2**  Let $x, y \in H$. Then $(xy)^2 = xyxy$, and, since $G$ is abelian, $xyxy = x^2 y^2 = ee = e$, so $(xy)^2 = e$ and therefore $xy \in H$. Since $e^2 = e$, $e \in H$. For $x \in H$, $x^2 = e$, so $x^{-2} x^2 = x^{-2} e$, or $e = x^{-2}$, and $x^{-1} \in H$. Therefore, by Theorem 21, $H$ is a subgroup of $G$.

**3**  Let $x, y \in H$. Then $(xy)^3 = xyxyxy$, and, since $G$ is abelian, $xyxyxy = x^3 y^3 = e^3 = e$, so $(xy)^3 = e$ and therefore $xy \in H$. Since $e^3 = e$, $e \in H$. For $x \in H$, $x^3 = e$, so $x^{-3} x^3 = x^{-3} e$, or $e = x^{-3}$, and $x^{-1} \in H$. Therefore, by Theorem 21, $H$ is a subgroup of $G$.

**4**  Let $x, y \in H$. Suppose $x$ has order $m$, and $y$ has order $n$. Then, as $G$ is abelian, $(xy)^{mn} = x^{mn} y^{mn} = \left(x^m\right)^n \left(y^n\right)^m = e^n e^m = e$. Therefore $xy$ has finite order, so $xy \in H$. As $e^1 = e$, the order of $e$ is finite, so

$e \in H$. For $x \in H$, if $x$ has finite order $m$, $x^m = e$. Then $x^{-m}x^m = x^{-m}e$ and $e = x^{-m}$, and therefore $\left(x^{-1}\right)^m = e$, and $x^{-1}$ has finite order. So $x^{-1} \in H$. Thus, by Theorem 21, $H$ is a subgroup of $G$.

**5**   (1) T. (2) F. See Example 5.2.2. (3) F. In $(\mathbf{C}^*, \times)$, the order of $i$ is 4. (4) F. The group of invertible $2 \times 2$ matrices under multiplication is an infinite group, and the matrices which have determinant 1 form a non-abelian subgroup. (5) F. The trivial group consisting of $\{e\}$ alone, has no proper subgroups. More significantly, the group $\mathbf{Z}_p$, where $p$ is prime, has no proper subgroups.

**6**   Let $x, y \in A \cap B$. Then $x, y \in A$ and $x, y \in B$, and as $A$ and $B$ are subgroups, $xy \in A$ and $xy \in B$. So $xy \in A \cap B$. As $A$ and $B$ are both subgroups, $e \in A$ and $e \in B$. So $e \in A \cap B$. Let $x \in A \cap B$. Then $x \in A$ and $x \in B$. So, as $A$ and $B$ are subgroups, $x^{-1} \in A$ and $x^{-1} \in B$, so that $x^{-1} \in A \cap B$. So, by Theorem 21, $A \cap B$ is a subgroup of $G$.

It is not true that $A \cup B$ is a subgroup of $G$. Consider the group $D_3$ in Fig. 5.2. $\{I, X\}$ and $\{I, Y\}$ are subgroups, but $\{I, X, Y\}$ is not.

**7**   Since $a \in H$, and $H$ is a subgroup, $a^2 = aa \in H$. And since $a \in H$ and $a^2 \in H$, $a^3 = aa^2 \in H$. It follows, formally by induction, that $a^n \in H$ for $n \in \mathbf{Z}^+$. Similarly, since $a \in H$ and $H$ is a subgroup, $a^{-1} \in H$, and therefore $a^{-m} \in H$ for $m \in \mathbf{Z}^+$. And finally $e = a^0 \in H$ because $H$ is a subgroup. Therefore, since $\langle a \rangle = \left\{a^n : n \in \mathbf{Z}\right\}$, it follows that $\langle a \rangle \subseteq H$.

**8** *If.* Suppose that $ab^{-1} \in H$ for all $a, b \in H$. Since $H$ is non-empty, there exists an element $x \in H$. Therefore $xx^{-1} \in H$, and so $e \in H$. And since $e, x \in H$, $x^{-1} = ex^{-1} \in H$. Finally, for any $x, y \in H$, as $y^{-1} \in H$, $xy = x\left(y^{-1}\right)^{-1} \in H$. Therefore, by Theorem 21, $H$ is a subgroup of $G$.

*Only if.* Suppose that $H$ is a subgroup of $G$. Then $H$ is non-empty, since $e \in H$. Then for any element $y \in H$, its inverse, $y^{-1} \in H$. Thus, if $x, y \in H$, then $x, y^{-1} \in H$, and, by Theorem 21, $xy^{-1} \in H$.

**9**   Let $x, y \in H$. Then $gxy = xgy = xyg$, as $gx = xg$ and $gy = yg$. Therefore $g(xy) = (xy)g$, so $xy \in H$. As $ge = g = eg$, $e \in H$. Finally, let $x \in H$. Then, as $gx = xg$, $x^{-1}gxx^{-1} = x^{-1}xgx^{-1}$, so $x^{-1}ge = egx^{-1}$ and $x^{-1}g = gx^{-1}$. Therefore $x^{-1} \in H$. Therefore, by Theorem 21, $H$ is a subgroup of $G$.

**10**   As $H$ is a subset of $K$, and $K$ is a subset of $G$, $H$ is a subset of $G$. Then, as $H$ is a subgroup of $K$, $H$ is a group under the operation of $K$; and as $K$ is a subgroup of $G$, $K$ is a group under the operation of $G$. So, $H$ is a group under the operation of $G$. Therefore $H$ is a subgroup of $G$.

## CHAPTER 7

**1**   Show that 2 (or 8) is a generator. $2^1 = 2$, $2^2 = 4$, $2^3 = 8$ and $2^4 = 6$, so 2 is a generator for $(\{2,4,6,8\}, \times \bmod 10)$, which is therefore cyclic.

**2**   Proper subgroups of a cyclic group of order 12 with generator $g$, are $\{e, g^6\}$, $\{e, g^4, g^8\}$, $\{e, g^3, g^6, g^9\}$ and $\{e, g^2, g^4, g^6, g^8, g^{10}\}$.
These subgroups are $\langle g^6 \rangle$, $\langle g^4 \rangle$, $\langle g^3 \rangle$ and $\langle g^2 \rangle$.

*If you look forward to Theorem 52, you will see that these are the only proper subgroups.*

**3**   1 and 5.

**4**   (1) T. (2) F. In $\mathbf{Z}_4$, 1 and 3 are generators, but 2 is not. (3) T. See part (2). (4) F. See Theorem 24. (5) T. See Theorem 23. (6) F. See Example 7.2.3. (7) F. For suppose $z$ is a generator. If $|z| \neq 1$, no power of $z$ can equal $i$. If $|z| = 1$, no power of $z$ can equal 2.

**5**   For any non-zero element $g \in \mathbf{R}$, $\frac{1}{2}g$ is an element of $\mathbf{R}$ which is not an integer multiple of $g$. Therefore $g$ is not a generator. Therefore $(\mathbf{R}, +)$ has no generator and so is not cyclic.

*Remember that when you use additive notation for a group, $g^n$ becomes $ng$.*

**6**   (1) Not cyclic. $5^2 = 1$, $7^2 = 1$ and $11^2 = 1$ so there is no generator. (2) Not cyclic. The argument is the same as for question 5. (3)   Not cyclic. First note that an infinite cyclic group cannot have an element of finite order, except for the identity. For if $a$ is a generator of the infinite cyclic group, and $b \, (\neq e)$ is a element of finite order. Then $b^d = e$ for some positive integer $d$, and, since $a$ is a generator, $b = a^n$ for some positive integer $n$. Therefore $a^{nd} = e$, $a$ has finite order and the group can not be infinite. But $T$ is infinite, and does have elements of finite order other than the identity; for example, $-1$ has order 2. Therefore $T$ is not a cyclic group.

(4)   Not cyclic. Suppose that $a + bi$ is a generator. Then as $i \in \mathbf{Z}[i]$, $i = n(a + bi)$ for some integer $n$. So, equating real and imaginary parts, $na = 0$ and $nb = 1$. The only solutions are $n = 1$, $b = 1$ and $a = 0$, giving $i$, or $n = -1$, $b = -1$ and $a = 0$, giving $-i$. But neither $i$ nor $-i$ is a generator, as no number of the form $ni$ or $n(-i)$ can equal 2.

## CHAPTER 8

**1**   (1) F. Think of $\mathbf{Z} \times \mathbf{Z}$. (2) T. (3) T. (4) F.

**2**   $A \times B \times C = \{(a,b,c) : a \in A, b \in B, c \in C\}$.        $\mathbf{Z}_2 \times \mathbf{Z}_2 \times \mathbf{Z}_2 = \{(0,0,0),(0,0,1),(0,1,0),(0,1,1),(1,0,0),(1,0,1),(1,1,0),(1,1,1)\}$.

**3** Suppose that $(a,b) \in \mathbf{Z} \times \mathbf{Z}$. Then $a \in \mathbf{Z}$ and $b \in \mathbf{Z}$. But $\mathbf{Z} \subseteq \mathbf{Q}$. Therefore $a \in \mathbf{Q}$ and $b \in \mathbf{Q}$, so $(a,b) \in \mathbf{Q} \times \mathbf{Q}$. Therefore $\mathbf{Z} \times \mathbf{Z} \subseteq \mathbf{Q} \times \mathbf{Q}$.

*This is a particular case of a more general situation. If $A \subseteq S$ and $B \subseteq T$, then $A \times B \subseteq S \times T$.*

**4** You can interpret $\mathbf{Z} \times \mathbf{Z}$ as an integer grid of dots on the infinite sheet of paper $\mathbf{R} \times \mathbf{R}$.

**5**

|       | (0,0) | (0,1) | (1,0) | (1,1) |
|-------|-------|-------|-------|-------|
| (0,0) | (0,0) | (0,1) | (1,0) | (1,1) |
| (0,1) | (0,1) | (0,0) | (1,1) | (1,0) |
| (1,0) | (1,0) | (1,1) | (0,0) | (0,1) |
| (1,1) | (1,1) | (1,0) | (0,1) | (0,0) |

**6** Let $g_1, g_2 \in G$ and $h_1, h_2 \in H$. As $G, H$ are abelian, $g_1 g_2 = g_2 g_1$ and $h_1 h_2 = h_2 h_1$. Now consider $(g_1, h_1)(g_2, h_2) = (g_1 g_2, h_1 h_2)$. $= (g_2 g_1, h_2 h_1) = (g_2, h_2)(g_1, h_1)$, showing that $G \times H$ is abelian.

**7** Since $G$ is not abelian, there exist elements $g_1, g_2 \in G$ such that $g_1 g_2 \neq g_2 g_1$. Let $h_1, h_2 \in H$. Then $(g_1, h_1)(g_2, h_2) = (g_1 g_2, h_1 h_2)$ and $(g_2, h_2)(g_1, h_1) = (g_2 g_1, h_2 h_1)$, and since $g_1 g_2 \neq g_2 g_1$, the elements $(g_1 g_2, h_1 h_2) \neq (g_2 g_1, h_2 h_1)$. Therefore $(g_1, h_1)(g_2, h_2) \neq (g_2, h_2)(g_1, h_1)$ showing that $G \times H$ is not abelian.

**8**

|       | (0,0) | (0,1) | (0,2) | (1,0) | (1,1) | (1,2) |
|-------|-------|-------|-------|-------|-------|-------|
| (0,0) | (0,0) | (0,1) | (0,2) | (1,0) | (1,1) | (1,2) |
| (0,1) | (0,1) | (0,2) | (0,0) | (1,1) | (1,2) | (1,0) |
| (0,2) | (0,2) | (0,0) | (0,1) | (1,2) | (1,0) | (1,1) |
| (1,0) | (1,0) | (1,1) | (1,2) | (0,0) | (0,1) | (0,2) |
| (1,1) | (1,1) | (1,2) | (1,0) | (0,1) | (0,2) | (0,0) |
| (1,2) | (1,2) | (1,0) | (1,1) | (0,2) | (0,0) | (0,1) |

Strictly $\mathbf{Z}_2 \times \mathbf{Z}_3$ is not the same group as $\mathbf{Z}_3 \times \mathbf{Z}_2$.

**9** $(1,1)^2 = (2,0)$, $(1,1)^3 = (0,1)$, $(1,1)^4 = (1,0)$, $(1,1)^5 = (2,1)$ and $(1,1)^6 = (0,0)$. You can check that $(2,1)$ is also a generator. All the other elements of $\mathbf{Z}_3 \times \mathbf{Z}_2$ have at least one 0, so powers of these elements will also have at least one 0. So no element of $\mathbf{Z}_3 \times \mathbf{Z}_2$, other than $(1,1)$ and $(2,1)$, can be a generator.

**10** In $G \times H$, $(x,y)^n = (x^n, y^n)$, and as $(x,y)$ has order $n$, $(x,y)^n = (e_G, e_H)$. Then $(x^n, y^n) = (e_G, e_H)$, so $x^n = e_G$ and $y^n = e_H$. So the order of $x$ in $G$ divides $n$, and the order of $y$ in $H$ divides $n$.

## CHAPTER 9

**1** As $f(0) = f(\pi)$, $f$ is not an injection. And as there is no element $x \in \mathbf{R}$ such that $\sin x = 2$, $f$ is not a surjection.

**2**     $f(x) = e^x$ is an example of a function $f : \mathbf{R} \to \mathbf{R}$ which is an injection but not a surjection. It is an injection because if $f(x) = f(y)$, then $e^x = e^y$, and, taking natural logarithms, $x = y$. However, $f$ is not a surjection since there is no element $x$ such that $f(x) = 0$.

**3**     $f(x) = x^3 - x$ is an example of a function $f : \mathbf{R} \to \mathbf{R}$ which is a surjection but not an injection. It is surjective, since it is always possible to solve the equation $f(x) = c$, or $x^3 - x = c$, for $c \in \mathbf{R}$. (Think of the graph of a cubic.) Or alternatively, the expression $x^3 - x - c$ can be factored into a product of a quadratic and a linear function. Therefore there is bound to be one real root, and possibly three, of the equation $x^3 - x - c = 0$. However, $f$ is not an injection since $(1)^3 - (1) = 0 = (0)^3 - (0)$, so $f(0) = f(1)$.

**4**     (1) Function; not an injection as $f(-1) = f(1)$; not a surjection as no element maps to $-1$. (2) Function; *injection*. Suppose that $f(x) = f(y)$. Then $x^3 = y^3$ so $x^3 - y^3 = (x - y)(x^2 + xy + y^2) = 0$. So either $x = y$ or $x^2 + xy + y^2 = 0$. But $x^2 + xy + y^2 = \left(x + \frac{1}{2}y\right)^2 + \frac{3}{4}y^2$, which is the sum of two squares, and is only zero when both brackets are zero, i.e., when $x = y = 0$. So in both cases, $x = y$. *surjection*. For $x \in \mathbf{R}$, $x^{1/3} \in \mathbf{R}$ and $f\left(x^{1/3}\right) = x$, so $f$ is a surjection. (3) Not a function, because it is not defined at $x = 0$. (4) Function; not an injection as $f(0) = f(2\pi)$; not a surjection as no element maps to 2. (5) Not a function, because it is not defined at $x = \frac{1}{2}\pi$. (6) Function; See the answer to question 2. (7) Function; not an injection, since $f(-1) = f(1)$, not a surjection as no element maps to $-1$. (8) Function; not an injection as $f(1.1) = f(1.2) = 1$; not a surjection as no element maps to 0.5. (9) Not a function, because it is not defined for $-1$. (10) Function; *injection*, since if $f(x) = f(y)$, $x + 1 = y + 1$ then $x = y$; *surjection*; since for $a \in \mathbf{R}$, $f(a - 1) = a$. (11) Not a function, as it is not defined for 2. (12) Not a function; there is no smallest real number greater than a given real number.

**5**     *Injection.* Suppose that $f(x) = f(y)$. Then $\sqrt{x} = \sqrt{y}$. Squaring gives $x = y$, so that $f$ is an injection.

*Surjection.* If $y$ is any element of the co-domain, then $f\left(y^2\right) = y$.

So, as $f$ is both an injection and a surjection, it is a bijection.

**6**     *Injection.* Suppose that $f(x) = f(y)$. Then $1/x = 1/y$, so $x = y$.

*Surjection.* If $y$ is any element of the co-domain, then as $y \neq 0$, $1/y$ is an element of the domain and $f(1/y) = y$. So $f$ is a surjection.

Therefore, as $f$ is both an injection and a surjection, it is a bijection.

**7**   *Surjection*: if $m \in \mathbf{Z}^+$ is even, $m = 2n$ where $n > 0$ and for that $n$, $f(n) = m$; if $m \in \mathbf{Z}^+$ is odd, $m - 1 \geq 0$ and is even, so $(1 - m)/2$ is an integer $n \leq 0$. For that $n$, $f(n) = m$. *Injection*: because if $f(n) = f(m)$ where $m$ and $n$ are both $> 0$, $2m = 2n$ and $m = n$. If $f(n) = f(m)$ where $m$ and $n$ are both $\leq 0$, $1 - 2m = 1 - 2n$ and $m = n$. And if $f(n) = f(m)$ where one of them, $n$ say, is $> 0$, and the other is $\leq 0$, $1 - 2m = 2n$. But the left-hand side is even and the right-hand side is odd, an impossibility. Therefore, since $f$ is a surjection and an injection, $f$ is a bijection.

**8**   (1) You should realise here that assumptions are being made that the function is a function $f : \mathbf{R} \to \mathbf{R}$, and that a graph can actually be drawn in the high school sense. However, with these assumptions, a function is an injection if every line drawn parallel to the $x$-axis intersects the graph at most once. (2) A function is a surjection if every line drawn parallel to the $x$-axis meets the graph at least once. (3) A function is a bijection if every line drawn parallel to the $x$-axis meets the graph exactly once.

**9**   (1) F. $f$ is not an injection since $f(0) = f(1)$, neither is $f$ a surjection since there is no element which maps to 1. (2) T. (3) T. See definition. (4) T. $f$ is an injection since if $f(x, y) = f(p, q)$, $(y, x) = (q, p)$, so $y = q$ and $x = p$. And $f$ is a surjection since, given $(p, q) \in \mathbf{R} \times \mathbf{R}$, $(q, p)$ maps to it. (5) F. $f$ is not defined for $x = 0$.

**10**   (1) Not an injection, because if $p_1(x) = x$ and $p_2(x) = x + 1$, then $f(p_1) = f(p_2)$; surjection, because, given a polynomial $p(x)$,

$$f\left(\int_0^x p(t)\,dt\right) = p(x).$$ (2) Not a function. Because of the constant of integration, $f(p)$ is not uniquely specified. (3) Injection, because if $f(p_1) = f(p_2)$, then $\int_0^x p_1(t)\,dt = \int_0^x p_2(t)\,dt$. Differentiating gives $p_1 = p_2$. Not a surjection because no function integrates to give the polynomial 1. (4) Injection; if $f(p_1) = f(p_2)$, then $xp_1(x) = xp_2(x)$, so $p_1(x) = p_2(x)$ or $p_1 = p_2$. Not a surjection because no polynomial $p$ has the property that $f(p) = xp(x) = 1$.

**11**   *Surjection*. Let $y \in G$. Then $f\left(yg^{-1}\right) = \left(yg^{-1}\right)g = y\left(gg^{-1}\right) = ye$ $= y$, so $f$ is a surjection. *Injection*. If $f(x) = f(y)$, then $xg = yg$, and, using the cancellation rule, Theorem 15, part (4), $x = y$, so $f$ is an injection. Therefore $f$ is a bijection.

This function shows that, for a finite group, all the elements in a column of the group table are different. A similar bijection, $f : G \to G$, such that $f(x) = gx$ shows the same for rows.

## CHAPTER 10

**1**   Given $y \in A$, $y = I_A(y) = g(f(y))$ so $g$ is a surjection.
If $f(x) = f(y)$, then $g(f(x)) = g(f(y))$, so $(g \circ f)(x) = (g \circ f)(y)$. Therefore $I_A(x) = I_A(y)$, so $x = y$. Therefore $f$ is an injection.

**2**   The only bijections are (2) and (10). (2) The inverse is $f^{-1} : \mathbf{R} \to \mathbf{R}$ where $f^{-1}(x) = x^{\frac{1}{3}}$. (10) The inverse is $f^{-1} : \mathbf{R} \to \mathbf{R}$ where $f^{-1}(x) = x - 1$.

**3**   Consider first $f \circ g : \mathbf{Z}^+ \to \mathbf{Z}^+$. For $n$ even, $f(g(n)) = f(n/2) = 2(n/2) = n$. For $n$ odd, $f(g(n)) = f((1-n)/2) = 1 - 2(1-n)/2 = n$.
Now consider $g \circ f : \mathbf{Z} \to \mathbf{Z}$. For $n > 0$, $g(f(n)) = g(2n) = 2n/2 = n$. For $n \le 0$, $g(f(n)) = g(1 - 2n) = (1 - (1 - 2n))/2 = n$.
So, by Theorem 30, and the definition which follows, $g : \mathbf{Z}^+ \to \mathbf{Z}$ is the inverse of $f : \mathbf{Z} \to \mathbf{Z}^+$.

This is a quicker method of proving that $f : \mathbf{Z} \to \mathbf{Z}^+$ is a bijection.

**4**   (1) F. Should be $g^{-1} \circ f^{-1}$. (2) T. See definition. (3) T. See definition. (4) T. See Theorem 27. (5) F. $f : \mathbf{Z} \to \mathbf{Z}$ defined by $f(n) = 2n$ is an injection, but not a surjection.

**5**   Let $f, g \in H$. Then $(f \circ g)(X) = f(g(X)) = f(X) = X$ so $f \circ g \in H$ and $H$ is closed. Since $e_B(X) = X$, $e_B \in H$. Let $f \in H$ and let $f^{-1}$ be the inverse in $B$ of $f$. Then $f^{-1}(X) = f^{-1}(f(X)) = (f^{-1} \circ f)(X) = e_B(X) = X$ so $f^{-1} \in H$. So $H$ is a subgroup of $B$.

## CHAPTER 11

**1**   $\mathbf{Z}_2 \times \mathbf{Z}_2$ is not isomorphic to $\mathbf{Z}_4$. In $\mathbf{Z}_2 \times \mathbf{Z}_2$ every element, apart from the identity, has order 2, while $\mathbf{Z}_4$ is cyclic with an element of order 4.

**2**   $\mathbf{Z}_2 \times \mathbf{Z}_2$ is isomorphic to $V$. The bijection $f : \mathbf{Z}_2 \times \mathbf{Z}_2 \to V$ given by $f((0,0)) = I$, $f((1,0)) = X$, $f((0,1)) = Y$ and $f((1,1)) = H$ is an isomorphism; check by considering each pair such as $f((0,1) + (1,1)) = f((1,0)) = X$ and $f((0,1))f((1,1)) = YH = X$ and showing that you get the same result.

**3**   If two groups $G$ and $H$ are isomorphic, you can show that an element $g \in G$ is a generator of $G$ if, and only if, its image $f(g) = h$ is a generator of $H$.

Notice that this actually follows from the result of question 5.

There are only two generators of $\mathbf{Z}_6$, namely 1 and 5. Either of these could be mapped to $(1,1)$, giving two isomorphisms.

**4** *Injection.* $f$ is an injection since if $f(x) = f(y)$, then $g^{-1}xg = g^{-1}yg$. Therefore $g(g^{-1}xg)g^{-1} = g(g^{-1}yg)g^{-1}$ leading to $x = y$.
*Surjection.* Take $x \in G$, and consider the element $gxg^{-1} \in G$. Then $f(gxg^{-1}) = g^{-1}(gxg^{-1})g = x$.
Finally, $f(xy) = gxyg^{-1} = gxg^{-1}gyg^{-1} = f(x)f(y)$, so $f : G \rightarrow G$ is an isomorphism.

**5** Since the order of $g$ is $n$, $g^n = e_G$, so $f(g^n) = f(e_G) = e_H$. But $f(g^n) = f(g)^n = h^n$, so $h^n = e_H$. Suppose that $h^r = e_H$. Then $f(g^r) = f(g)^r = h^r = e_H = f(e_G)$, and so $g^r = e_G$ because $f$ is an injection. Therefore $r \geq n$, because $n$ is the order of $g$. Therefore $n$ is the least positive integer such that $h^n = e_H$; that is, the order of $h$ is $n$.

*A good way to show that two groups are not isomorphic can be to show that the numbers of elements of the same order in the two groups are different. This is related to Section 11.4 which involved the solutions of the equation $x^n = e$ in each group.*

**6** Define $f : \mathbf{R} \rightarrow \mathbf{R}^+$ by $f(x) = e^x$. This is a bijection since, using Theorem 30, $g : \mathbf{R}^+ \rightarrow \mathbf{R}$ such that $g(x) = \ln x$ has the property that $\ln(e^x) = x$ for all $x \in \mathbf{R}$, and $e^{\ln x} = x$ for all $x \in \mathbf{R}^+$. And since $f(x + y) = e^{x+y} = e^x \times e^y = f(x)f(y)$, $f : \mathbf{R} \rightarrow \mathbf{R}^+$ is an isomorphism.

**7** Define $f : \mathbf{Z} \rightarrow 3\mathbf{Z}$ by $f(n) = 3n$. Injection, for if $f(m) = f(n)$, $3m = 3n$, so $m = n$. Surjection, because if $m \in 3\mathbf{Z}$, $m = 3n$ for some $n \in \mathbf{Z}$, and for that $n$, $f(n) = 3n = m$. And finally, $f(m + n) = 3(m + n) = 3m + 3n = f(m) + f(n)$. Therefore $f : \mathbf{Z} \rightarrow 3\mathbf{Z}$ is an isomorphism, and the groups are isomorphic.

**8** Let $G$ be a non-abelian group. Then there exist $x, y \in G$ such that $xy \neq yx$. Let $H$ be an abelian group. Suppose that there is an isomorphism $f : G \rightarrow H$ such that $f(x) = X$ and $f(y) = Y$. Then $f(xy) = f(x)f(y) = XY = YX = f(y)f(x) = f(yx)$ so that $xy$ and $yx$, which are different, have the same image. But, as $f$ is an isomorphism, it is also a bijection, and hence an injection, so different elements have different images. So the supposition is false. No isomorphism exists.

**9** Define $f : G \times H \rightarrow H \times G$ by $f(g, h) = (h, g)$. Injection, for if $f(g_1, h_1) = f(g_2, h_2)$, then $(h_1, g_1) = (h_2, g_2)$, so $h_1 = h_2$ and $g_1 = g_2$, which leads to $(g_1, h_1) = (g_2, h_2)$. Surjection, for if $(h, g)$ is any element of $H \times G$, then $f(g, h) = (h, g)$. Finally, $f(g_1, h_1)f(g_2, h_2) = (h_1, g_1)(h_2, g_2) = (h_1 h_2, g_1 g_2)$ and $f((g_1, h_1)(g_2, h_2)) = f(g_1 g_2, h_1 h_2) = (h_1 h_2, g_1 g_2)$, so $f(g_1, h_1)f(g_2, h_2) = f((g_1, h_1)(g_2, h_2))$. Therefore $f$ is an isomorphism, and $G \times H \cong H \times G$.

**10**    Define $f : \mathbf{Z}_{mn} \to \mathbf{Z}_m \times \mathbf{Z}_n$ by $f\left([a]_{mn}\right) = \left([a]_m, [a]_n\right)$.

*Well defined.* If $[a]_{mn} = [b]_{mn}$ then $a \equiv b \,(\mathrm{mod}\, mn)$, so $mn$ divides $a - b$. Therefore $m$ divides $a - b$ and $n$ divides $a - b$. It follows that $a \equiv b \,(\mathrm{mod}\, m)$ and $a \equiv b \,(\mathrm{mod}\, n)$ so $[a]_m = [b]_m$ and $[a]_n = [b]_n$. Therefore $\left([a]_m, [a]_n\right) = \left([b]_m, [b]_n\right)$.

*Injection.* If $f\left([a]_{mn}\right) = f\left([b]_{mn}\right)$, then $\left([a]_m, [a]_n\right) = \left([b]_m, [b]_n\right)$ so $[a]_m = [b]_m$ and $[a]_n = [b]_n$. Thus $a \equiv b \,(\mathrm{mod}\, m)$ and $a \equiv b \,(\mathrm{mod}\, n)$. But as $m$ and $n$ are relatively prime, $mn$ divides $a - b$, from Theorem 3. Therefore $a \equiv b \,(\mathrm{mod}\, mn)$ and $[a]_{mn} = [b]_{mn}$.

*Surjection.* $\mathbf{Z}_{mn}$ and $\mathbf{Z}_m \times \mathbf{Z}_n$ have the same number of elements, so by Theorem 27, $f$ is a surjection.

Finally : $f\left([a]_{mn} + [b]_{mn}\right) = f\left([a + b]_{mn}\right)$

$$= \left([a + b]_m, [a + b]_n\right)$$

$$= \left([a]_m + [b]_m, [a]_n + [b]_n\right)$$

$$= \left([a]_m, [a]_n\right) + \left([b]_m, [b]_n\right)$$

$$= f\left([a]_{mn}\right) + f\left([b]_{mn}\right)$$

Therefore $f$ is an isomorphism.

**11**    In the solution to question 10, it was proved that $f : \mathbf{Z}_{mn} \to \mathbf{Z}_m \times \mathbf{Z}_n$ defined by $f\left([a]_{mn}\right) = \left([a]_m, [a]_n\right)$ is surjective. This means that, for any $[a]_m \in \mathbf{Z}_m$ and $[b]_n \in \mathbf{Z}_n$, there is an element $[x]_{mn} \in \mathbf{Z}_{mn}$ such that $f\left([x]_{mn}\right) = \left([a]_m, [b]_n\right)$. It follows that $\left([x]_m, [x]_n\right) = \left([a]_m, [b]_n\right)$, so $[x]_m = [a]_m$ and $[x]_n = [b]_n$, so $x \equiv a \,(\mathrm{mod}\, m)$ and $x \equiv b \,(\mathrm{mod}\, n)$.

**12**    Let $G$ and $G'$ be isomorphic groups, and let $H$ be a subgroup of $G$. Let $f : G \to G'$ be an isomorphism, and let $H' = \left\{ f(h) : h \in H \right\}$, be the image of $H$ in $G'$. Let $x, y \in H'$. Then, from the definition of $H'$, there exist $a, b \in H$ such that $f(a) = x$ and $f(b) = y$. Then $xy = f(a)f(b) = f(ab)$. But since $H$ is a subgroup, $ab \in H$, and, by the definition of $H'$, $f(ab) \in H'$. Therefore $xy \in H'$. Since $e \in H$, $f(e)$, which is the identity in $G'$, is a member of $H'$. Finally, suppose that $x = f(a) \in H'$. Then, since $H$ is a subgroup of $G$, $a^{-1} \in H$ so $f\left(a^{-1}\right) \in H'$. But $f\left(a^{-1}\right) = \left(f(a)\right)^{-1}$, from Theorem 15, so $\left(f(a)\right)^{-1} = x^{-1} \in H'$. So $H'$ is a subgroup of $G'$.

*Notice that the bijectivity of $f$ was not used.*

**13**    Let $f : G \to H$ be an isomorphism. Let $g$ be a generator for the cyclic group $G$, and let $f(g) = h$. As a preliminary, we need to prove that $f(g^r) = h^r$ for all $r \in \mathbf{Z}$. For any positive integer $r$,

$$f(g^r) = \overbrace{f(g)f(g)\dots f(g)}^{r \text{ times}} = h^r.$$ (You can prove this formally by induction.) Suppose now that $s$ is a negative integer, so that $-s$ is positive. Then $f(g^{-s}) = h^{-s}$. Therefore $f(g^s)f(g^{-s}) = f(g^s)h^{-s}$. But $f(g^s)f(g^{-s}) = f(g^s g^{-s}) = f(e_G) = e_H$, so $f(g^s)h^{-s} = e_H$. Therefore $f(g^s)$ is the inverse of $h^{-s}$, so it follows that $f(g^s) = h^s$. Finally, $f(g^0) = f(e_G) = e_H = h^0$, so $f(g^r) = h^r$ for all $r \in \mathbf{Z}$.

Given $y \in H$, there exists $x \in G$ such that $f(x) = y$, since $f$ is surjective. As $G = \langle g \rangle$, $x = g^n$ for some $n \in \mathbf{Z}$. So $y = f(x) = f(g^n) = h^n$. Therefore $h$ is a generator for $H$, so $H = \langle h \rangle$.

*Notice that the injectivity of f was not used.*

**14** Let $f, g \in A$. Then $f$ and $g$ are bijections. By Theorem 34, the bijections form a group. It only remains to prove that the set of isomorphisms is a subgroup of the group of bijections. As $(fg)(xy) = f(g(xy)) = f(g(x)g(y)) = f(g(x))f(g(y)) = (fg)(x)(fg)(y)$, the composite $fg$ is an isomorphism. The identity function $I$ is an isomorphism, because $I(xy) = xy = I(x)I(y)$. Finally suppose $f(x) = X$ and $f(y) = Y$. Then $f^{-1}(XY) = f^{-1}(f(x)f(y)) = f^{-1}(f(xy)) = (f^{-1}f)(xy) = xy = f^{-1}(X)f^{-1}(Y)$, so $f^{-1}$ is an isomorphism. Therefore, by Theorem 21, the group of isomorphisms is a subgroup of the group of bijections, and so a group.

**15** $f(xy) = (xy)^s = x^s y^s = f(x)f(y).$

**16** *If.* Suppose that $s = \pm 1$. Then $f(a) = \pm a$ for all $a \in \mathbf{Z}$. *Injection.* If $f(a) = f(b)$, $sa = sb$, so $a = b$ and $f$ is injective. *Surjection.* For $a \in \mathbf{Z}$, if $s = 1$ then $f(a) = a$, and if $s = -1$ then $f(-a) = a$, so $f$ is surjective. Therefore $f$ is bijective. Finally, using question 15, $f$ is an isomorphism. So if $s = \pm 1$, $f$ is an isomorphism.

*Only if.* If $f$ is an isomorphism, then $f$ is a bijection, and so $f(a) = 1$ for some $a$. Therefore $sa = 1$. The only solutions are $s = 1$, $a = 1$, or $s = -1$, $a = -1$. Therefore $s = \pm 1$.

**17** *If. Surjection.* As $s$ and $n$ are relatively prime, by Theorem 1, there exist integers $a$ and $b$ such that $na + sb = 1$. Therefore $sb \equiv 1 \pmod{n}$. Let $r \in \mathbf{Z}$. Then $rsb \equiv r \pmod{n}$, and it follows that $f([rb]) = [srb] = [r]$. Therefore $f$ is surjective. *Injection.* From

Theorem 27, $f$ is injective. Therefore $f$ is bijective. Finally, using the result of question 15, $f$ is an isomorphism.

*Only if.* As $f$ is an isomorphism, it is a bijection and hence a surjection. Therefore $f([a]) = [1]$ for some $[a] \in \mathbf{Z}_n$. Therefore $[sa] = [1]$, so $n$ divides $1 - as$, and $1 - as = bn$ for some integer $b$. Therefore $as + bn = 1$, so, by Theorem 1, $n$ and $s$ are relatively prime.

**18**  An automorphism of $(\mathbf{Z}, +)$ is an isomorphism of $(\mathbf{Z}, +)$ to itself. Suppose that $f \in Aut(\mathbf{Z})$ and that $f(1) = s$. Then, since $f$ is an isomorphism, $f(a) = \overbrace{f(1) + f(1) + \ldots + f(1)}^{a \text{ times}} = as$. From question 16, $s = \pm 1$, so $f$ must be one of the two functions $f_1$ and $f_2$, where $f_1(a) = a$ for all $a \in \mathbf{Z}$, and $f_2(a) = -a$ for all $a \in \mathbf{Z}$. But these are distinct and each is an automorphism of $\mathbf{Z}$. Therefore $Aut(\mathbf{Z}, +)$ is a group of order 2, and hence $Aut(\mathbf{Z}, +) \cong \mathbf{Z}_2$.

**19**  First part, to show that $U_n$ is a group under multiplication. To show that multiplication is closed on $U_n$, let $[s], [t] \in U_n$. Since $n$ and $s$ are relatively prime, by Theorem 1, there exist integers $a$ and $b$ such that $as + bn = 1$ and since $n$ and $t$ are relatively prime, there exist integers $c$ and $d$ such that $ct + dn = 1$. Multiplying these, $(ac)st + n(asd + btc + bdn) = 1$, so, by Theorem 1 again, $n$ and $st$ are relatively prime. Therefore $[st] \in U_n$.

*Associative.* The operation is associative: the proof follows the same lines as the corresponding part of the proof of Theorem 14.

*Identity.* As 1 and $n$ are relatively prime, $[1] \in U_n$. The identity element is $[1]$, because for any $[s] \in U_n$, $[s][1] = [s1] = [s]$ and $[1][s] = [1s] = [s]$.

*Inverse.* Finally, suppose that $[s] \in U_n$. Then, since $n$ and $s$ are relatively prime, there exist integers $a$ and $b$ such that $as + bn = 1$. Therefore $[a]$ is the inverse of $[s]$, because $[a][s] = [as] = [1 - bn] = [1]$, and $[s][a] = [sa] = [1 - bn] = [1]$. So $U_n$ is a group under multiplication.

For each integer $s$ which is relatively prime to $n$, let $f_s : \mathbf{Z}_n \to \mathbf{Z}_n$ be the function given by $f_s([a]) = [sa]$, for $[a] \in \mathbf{Z}_n$. By question 17, $f_s$ is an automorphism of $\mathbf{Z}_n$. Define $\phi : U_n \to Aut(\mathbf{Z}_n)$ by $\phi([s]) = f_s$, for all $[s] \in U_n$.

*Well defined.* If $[s] = [t]$, then $n$ divides $(s - t)$, so $n$ divides $a(s - t)$ for all $a$. Therefore $[sa] = [ta]$, for all $a$, so $f_s([a]) = f_t([a])$ for all $[a]$. Therefore $f_s = f_t$.

*Injection.* If $\phi([s]) = \phi([t])$ then $f_s = f_t$, so $f_s([1]) = f_t([1])$. Therefore $[s] = [t]$.

*Surjection.* Let $f \in Aut(\mathbf{Z}_n)$. Then $f([1]) = [s]$ for some $[s] \in \mathbf{Z}_n$.

$$f([a]) = \overbrace{f([1]) + f([1]) + \ldots + f([1])}^{a \text{ times}} = [sa],$$ for

As $f$ is an isomorphism all $a$. Therefore, by question 17, $s$ and $n$ are relatively prime. Therefore $f = \phi([s])$, where $[s] \in U_n$.

Finally $f_{st}([a]) = [sta] = f_s([ta]) = f_s f_t([a])$, for all $a$. Therefore $f_{st} = f_s f_t$ and hence $\phi([s][t]) = \phi([st]) = f_{st} = f_s f_t = \phi([s])\phi([t])$.

Therefore $\phi$ is an isomorphism and $Aut(\mathbf{Z}_n) \cong U_n$.

**20**  $18 = 2 \times 3 \times 3$. From Theorem 39 the following groups are distinct: $\mathbf{Z}_2 \times \mathbf{Z}_3 \times \mathbf{Z}_3$ and $\mathbf{Z}_2 \times \mathbf{Z}_9$, and any abelian group of order 18 is isomorphic to one of them. $\mathbf{Z}_2 \times \mathbf{Z}_3 \times \mathbf{Z}_3 \cong \mathbf{Z}_3 \times \mathbf{Z}_6$ and $\mathbf{Z}_2 \times \mathbf{Z}_9 \cong \mathbf{Z}_{18}$. So there are two abelian groups of order 18, isomorphic to $\mathbf{Z}_3 \times \mathbf{Z}_6$ and $\mathbf{Z}_{18}$.

**21**  $36 = 2 \times 2 \times 3 \times 3$. From Theorem 39 the following groups are distinct: $\mathbf{Z}_2 \times \mathbf{Z}_2 \times \mathbf{Z}_3 \times \mathbf{Z}_3$, $\mathbf{Z}_4 \times \mathbf{Z}_3 \times \mathbf{Z}_3$, $\mathbf{Z}_2 \times \mathbf{Z}_2 \times \mathbf{Z}_9$ and $\mathbf{Z}_4 \times \mathbf{Z}_9$, and any abelian group of order 36 is isomorphic to one of them. $\mathbf{Z}_2 \times \mathbf{Z}_2 \times \mathbf{Z}_3 \times \mathbf{Z}_3 \cong \mathbf{Z}_6 \times \mathbf{Z}_6$, $\mathbf{Z}_4 \times \mathbf{Z}_3 \times \mathbf{Z}_3 \cong \mathbf{Z}_3 \times \mathbf{Z}_{12}$, $\mathbf{Z}_2 \times \mathbf{Z}_2 \times \mathbf{Z}_9 \cong \mathbf{Z}_2 \times \mathbf{Z}_{18}$, $\mathbf{Z}_4 \times \mathbf{Z}_9 \cong \mathbf{Z}_{36}$. So there are four groups of order 36, isomorphic to $\mathbf{Z}_6 \times \mathbf{Z}_6$, $\mathbf{Z}_3 \times \mathbf{Z}_{12}$, $\mathbf{Z}_2 \times \mathbf{Z}_{18}$ and $\mathbf{Z}_{36}$.

**22**  $180 = 2 \times 2 \times 3 \times 3 \times 5$. From Theorem 39 the following groups are distinct: $\mathbf{Z}_2 \times \mathbf{Z}_2 \times \mathbf{Z}_3 \times \mathbf{Z}_3 \times \mathbf{Z}_5$, $\mathbf{Z}_4 \times \mathbf{Z}_3 \times \mathbf{Z}_3 \times \mathbf{Z}_5$, $\mathbf{Z}_2 \times \mathbf{Z}_2 \times \mathbf{Z}_9 \times \mathbf{Z}_5$ and $\mathbf{Z}_4 \times \mathbf{Z}_9 \times \mathbf{Z}_5$, and any abelian group of order 180 is isomorphic to one of them. $\mathbf{Z}_2 \times \mathbf{Z}_2 \times \mathbf{Z}_3 \times \mathbf{Z}_3 \times \mathbf{Z}_5 \cong \mathbf{Z}_6 \times \mathbf{Z}_{30}$, $\mathbf{Z}_4 \times \mathbf{Z}_3 \times \mathbf{Z}_3 \times \mathbf{Z}_5 \cong \mathbf{Z}_3 \times \mathbf{Z}_{60}$, $\mathbf{Z}_2 \times \mathbf{Z}_2 \times \mathbf{Z}_9 \times \mathbf{Z}_5 \cong \mathbf{Z}_2 \times \mathbf{Z}_{90}$, and $\mathbf{Z}_4 \times \mathbf{Z}_9 \times \mathbf{Z}_5 \cong \mathbf{Z}_{180}$. So there are four groups of order 180, isomorphic to $\mathbf{Z}_6 \times \mathbf{Z}_{30}$, $\mathbf{Z}_3 \times \mathbf{Z}_{60}$, $\mathbf{Z}_2 \times \mathbf{Z}_{90}$ and $\mathbf{Z}_{180}$.

## CHAPTER 12

**1**  $ab = \begin{pmatrix} 1 & 2 & 3 & 4 & 5 \\ 3 & 4 & 1 & 2 & 5 \end{pmatrix}$     $ba = \begin{pmatrix} 1 & 2 & 3 & 4 & 5 \\ 1 & 4 & 5 & 2 & 3 \end{pmatrix}$

$a^2 b = \begin{pmatrix} 1 & 2 & 3 & 4 & 5 \\ 4 & 1 & 5 & 3 & 2 \end{pmatrix}$     $ac^{-1} = \begin{pmatrix} 1 & 2 & 3 & 4 & 5 \\ 2 & 4 & 3 & 1 & 5 \end{pmatrix}$

$(ac)^{-1} = \begin{pmatrix} 1 & 2 & 3 & 4 & 5 \\ 4 & 1 & 3 & 2 & 5 \end{pmatrix}$     $c^{-1}ac = \begin{pmatrix} 1 & 2 & 3 & 4 & 5 \\ 3 & 4 & 2 & 5 & 1 \end{pmatrix}$

**2**  $x = a^{-1}b = \begin{pmatrix} 1 & 2 & 3 & 4 & 5 \\ 5 & 2 & 3 & 1 & 4 \end{pmatrix}$  $x = a^{-1}cb^{-1} = \begin{pmatrix} 1 & 2 & 3 & 4 & 5 \\ 4 & 1 & 2 & 5 & 3 \end{pmatrix}$

**3**  8, 4, 8.

**4**   3, 5 and 2. Using Theorem 41, even, odd, odd.

**5**   Odd, odd.

**6**   $(135)(24)$, $(1342)$, $(14)(25)$.

**7**   (1) $\begin{pmatrix} 1 & 2 & 3 & 4 & 5 & 6 \\ 2 & 3 & 1 & 6 & 5 & 4 \end{pmatrix}$ (2) $\begin{pmatrix} 1 & 2 & 3 & 4 & 5 & 6 \\ 2 & 3 & 4 & 6 & 5 & 1 \end{pmatrix}$

(3) $\begin{pmatrix} 1 & 2 & 3 & 4 & 5 & 6 \\ 2 & 1 & 4 & 6 & 5 & 3 \end{pmatrix}$

**8**   (1) $(1473)$. (2) $(1465732)$. (3) $(14)(23)$. (4) $(123456)$.

**9**   Align the rectangle in the same way as in question 4 in Exercises 5. Label the vertices with the integer 1 in the top right-hand corner, and proceed clockwise with the integers 2, 3 and 4. Let $f$ be the function from the group of symmetries of the rectangle to $S_4$ defined by $f(\phi) = \begin{pmatrix} 1 & 2 & 3 & 4 \\ \phi(1) & \phi(2) & \phi(3) & \phi(4) \end{pmatrix}$. Then with the notation of question 4 in Exercises 5, $f(I) = e$, $f(X) = (12)(34)$, $f(Y) = (14)(23)$ and $f(R) = (13)(24)$.

|          | $e$      | $(12)(34)$ | $(14)(23)$ | $(13)(24)$ |
|----------|----------|------------|------------|------------|
| $e$      | $e$      | $(12)(34)$ | $(14)(23)$ | $(13)(24)$ |
| $(12)(34)$ | $(12)(34)$ | $e$      | $(13)(24)$ | $(14)(23)$ |
| $(14)(23)$ | $(14)(23)$ | $(13)(24)$ | $e$      | $(12)(34)$ |
| $(13)(24)$ | $(13)(24)$ | $(14)(23)$ | $(12)(34)$ | $e$      |

**10**   $\mathbf{Z}_4 \times S_3$ has two elements of order 3, namely $(0,(123))$ and $(0,(132))$. Moreover, these are the only such elements, for, using the result of Exercise 8, question 10, if $(x,y) \in \mathbf{Z}_4 \times S_3$ has order 3, then the order of $x$ in $\mathbf{Z}_4$ is 1 or 3, and hence 1, and the order of $y \in S_3$ is 1 or 3, and hence 3. On the other hand, $S_4$ has eight elements of order 3, $(123)$, $(132)$, $(124)$, $(142)$, $(134)$, $(143)$, $(234)$ and $(243)$. Similarly, $\mathbf{Z}_2 \times \mathbf{Z}_2 \times S_3$ has only two elements of order 3, $(0,0,(123))$ and $(0,0,(132))$. Therefore, using the result of Exercise 11, question 5, $S_4 \not\cong \mathbf{Z}_4 \times S_3$ and $S_4 \not\cong \mathbf{Z}_2 \times \mathbf{Z}_2 \times S_3$.

Alternatively, in $\mathbf{Z}_4 \times S_3$ there is an element $(1,(123))$ which has order 12. But there is no element of order 12 in $S_4$, so $S_4 \not\cong \mathbf{Z}_4 \times S_3$.

**11**   Let $x = (a_1 a_2 \ldots a_n)$ be of length $n$. Then $x^2 = (a_1 a_3 \ldots)$, $x^3 = (a_1 a_4 \ldots)$, $\ldots$, $x^{n-1} = (a_1 a_n \ldots)$ are all different, and $x^n = e$. Therefore $n$ is the smallest power of $x$ which gives the identity, so the order of $x$ is $n$.

**12** Let $a = a_n a_{n-1} \ldots a_1$ be a permutation made up of disjoint cycles of lengths $\alpha_1, \alpha_2, \ldots, \alpha_n$, and let $l = \mathrm{LCM}(\alpha_1, \alpha_2, \ldots, \alpha_n)$. Then $a^l = e$, because all the disjoint cycles commute, and the order of each of them divides $l$. Suppose that $a^r = e$. Then, since disjoint cycles commute, $a^r = a_n{}^r a_{n-1}{}^r \ldots a_1{}^r = e$. Suppose that $a_i{}^r \neq e$ for some $i \in \{1, 2, \ldots, n\}$. Then there is a symbol, $x$ say, such that $a_i{}^r(x) \neq x$. But since the cycles are disjoint, no other cycle affects $x$, so $a^r(x) \neq x$. This contradicts the fact that $a^r = e$. Therefore $a_i{}^r = e$ for all $i = 1, 2, \ldots, n$. Therefore $r$ is a multiple of each of $\alpha_1, \alpha_2, \ldots, \alpha_n$. Therefore $l \leq r$, and so the order of $a$ is $l$.

**13** From the result at the beginning of Section 12.6, every pemutation $x \in S_n$ can be written as a product of transpositions. But any transposition $(ab)$ can be written as a product of the form $(ab) = (1a)(1b)(1a)$. Combining these two results shows that every pemutation $x \in S_n$ can be written as a product of the transpositions $(12), (13), \ldots, (1n)$.

**14** Let $a = (1\,2\ldots n-1)$ and $b = (n-1\,n)$. By direct calculation you can see that $aba^{-1} = (1\,n)$, $a^2ba^{-2} = (2\,n)$, and, in general, proof by induction, $a^iba^{-i} = (i\,n)$. You can now use the result of question 13.

**15** *Well defined.* The function $f$ is well defined, because, since $x$, by virtue of being a member of $A_n$ is an even permutation, and $(12)$ is an odd permutation, their product, by Theorem 41, is an odd permutation, and therefore a member of $S_n - A_n$. The mapping is therefore well defined.

*Injection.* Suppose that $f(x) = f(y)$. Then $(12)x = (12)y$. Therefore, by multiplying by $(12)$, $x = y$, so $f$ is an injection.

*Surjection.* Let $y \in S_n - A_n$. Then $y$ is odd and therefore $(12)y$ is even. As $f((12)y) = (12)((12)y) = ((12)(12))y = y$, $f$ is a surjection.

Thus $f$ is a bijection, so $A_n$ and $S_n - A_n$ have the same number of elements. But $A_n$ and $S_n - A_n$ are disjoint, and their union is $S_n$, which has $n!$ elements. So the number of elements in $A_n$ is $\frac{1}{2}n!$.

# CHAPTER 13

**1** You can carry out a verification using a diagram. $(ba)^2 = (ba)(ba) = b(aba) = bb = e$. Also $\left(ba^i\right)^2 = \left(ba^i\right)\left(ba^i\right) = b\left(a^iba^i\right) = b\left(a^{i-1}(aba)a^{i-1}\right) = b\left(a^{i-1}ba^{i-1}\right) = \ldots = bb = e$.

**2**  (1) $a \circ ba = aba = b$; (2) $a^{-1} = a^{n-1}$; (3) From question 1, since $(ba)^2 = e$, $(ba)^{-1} = ba$. (4) $bab^{-1}a^{-1} = b(aba)a^{-1}a^{-1} = bba^{-2} = a^{n-2}$; (5) $ba \circ ba^2 = b(aba)a = bba = a$

**3**  The formal proof is by induction. Note that basis step is true for $i = 1$ because $ab = ab(aa^{-1}) = (aba)a^{-1} = ba^{-1} = ba^{n-1}$. Suppose that $a^i b = ba^{n-i}$ is true for $i = k$. Then $a^k b = ba^{n-k}$, and $a^{k+1}b = a(a^k b) = a(ba^{n-k}) = (ab)a^{n-k} = (ba^{-1})a^{n-k} = ba^{n-(k+1)}$. So, if the statement is true for $k$, it is also true for $k+1$. Therefore, by the principle of mathematical induction, the statement is true for all $i \geq 1$, and so for $i \in \{1, 2, 3, \ldots, n-1\}$.

**4**  The answer depends on whether $n$ is even or odd. When $n$ is odd, there are $n$ subgroups of order 2, all of the form $\{e, ba^i\}$ for $i = 0, 1, \ldots, n-1$. If $n$ is even, then add the subgroup $\{e, a^{n/2}\}$.

**5**  If $ba^i = ba^j$, then, by the cancellation rule, Theorem 15, part 4, $a^i = a^j$, and therefore $e = a^{j-i}$, so $j = i$. If $ba^i = a^j$, then $b = a^{j-i}$, so $b$ a power of $a$. This is a contradiction, so $ba^i \neq a^j$ for any $i$ or $j$.

**6**  There are eight elements which have order 3. These are the rotations $X$, $Y$, $Z$, $T$, $X^2$, $Y^2$, $Z^2$ and $T^2$. Note that $X^3 = Y^3 = Z^3 = T^3 = I$. The three half-turns, $A$, $B$ and $C$, about the mid-points of opposite edges, are each of order 2; $A^2 = B^2 = C^2 = I$. This gives the 12 rotational symmetries $I, A, B, C, X, Y, Z, T, X^2, Y^2, Z^2$ and $T^2$. The proper subgroups of $G$ are $\{I, A\}$, $\{I, B\}$, $\{I, C\}$, $\{I, A, B, C\}$, $\{I, X, X^2\}$, $\{I, Y, Y^2\}$, $\{I, Z, Z^2\}$ and $\{I, T, T^2\}$. Suppose that a subgroup contains $I$, $A$ and $X$. Then it also contains $X^2$, $AX = Z$, $XA = T$, and hence $Z^2$ and $T^2$. There are already more than six elements. In fact, it turns out that this subgroup must be the whole group. This is true for any other starting set of the type $I$, $A$ and $X$. Labelling the vertices 1, 2, 3 and 4 as in Fig. 13.7, let $f : G \rightarrow S_4$ be defined by $f(\phi) = \begin{pmatrix} 1 & 2 & 3 & 4 \\ \phi(1) & \phi(2) & \phi(3) & \phi(4) \end{pmatrix}$. Then $f(I) = e$, $f(A) = (12)(34)$, $f(B) = (13)(24)$, $f(C) = (14)(23)$, $f(X) = (234)$, $f(Y) = (143)$, $f(Z) = (124)$, $f(T) = (132)$, $f(X^2) = (243)$, $f(Y^2) = (134)$, $f(Z^2) = (142)$ and $f(T^2) = (123)$. All these permutations are even. As there are 12 of them, the order of $A_4$, they constitute the whole of $A_4$ and, as in Example 12.3.4, $G \cong A_4$.

**7**  Let $B = A \cap C_n \left(= \{e, a^s, a^{2s}, \ldots, a^{(d-1)s}\}\right)$. Then you need to show that $A = B \cup ba^m B$.

It is clear that $B \cup ba^m B \subseteq A$, since $B \subseteq A$ and $ba^m \in A$.

To prove that $A \subseteq B \cup ba^m B$, let $x \in A$. If $x \in B$, then $x \in B \cup ba^m B$, so suppose that $x \notin B$. Then $x \notin C_n$. Then $bx \in C_n$.

Therefore $a^{-m}bx \in C_n$, because $a^m \in C_n$. But $a^{-m}b = (ba^m)^{-1} \in A$.

Therefore $a^{-m}bx \in A$, because $x \in A$. Therefore $a^{-m}bx \in A \cap C_n = B$.

Therefore $x = (ba^m)(a^{-m}bx) \in ba^m B$.

## CHAPTER 14

**1** The left cosets of $\{e, a^2\}$ are $\{e, a^2\}$, $\{a, a^3\}$, $\{b, ba^2\}$, $\{ba, ba^3\}$. The right cosets are $\{e, a^2\}$, $\{a, a^3\}$, $\{b, ba^2\}$ and $\{ba, ba^3\}$. The left cosets of $\{e, b\}$ are $\{e, b\}$, $\{a^2, ba^2\}$, $\{a, ba^3\}$ and $\{a^3, ba\}$. The right cosets are $\{e, b\}$, $\{a^2, ba^2\}$, $\{a, ba\}$ and $\{a^3, ba^3\}$. In the first case, the left cosets and the right cosets are identical; in the second case they are not.

**2** First observe that the order of $\mathbf{Z}_2 \times \mathbf{Z}_3$ is 6, and the order of $\mathbf{Z}_2 \times \{0\}$ is 2, so that there are 3 cosets. The elements of $\mathbf{Z}_2 \times \{0\}$ are $(0,0)$ and $(1,0)$. The elements of $\mathbf{Z}_2 \times \mathbf{Z}_3$ are $(0,0)$, $(1,0)$, $(0,1)$, $(1,1)$, $(0,2)$ and $(1,2)$.

The cosets of $\mathbf{Z}_2 \times \{0\}$ are: $(0,0) + \mathbf{Z}_2 \times \{0\} = \mathbf{Z}_2 \times \{0\}$, $(1,0) + \mathbf{Z}_2 \times \{0\} = \mathbf{Z}_2 \times \{0\}$, $(0,1) + \mathbf{Z}_2 \times \{0\} = \mathbf{Z}_2 \times \{1\}$ $(1,1) + \mathbf{Z}_2 \times \{0\} = \mathbf{Z}_2 \times \{1\}$, $(0,2) + \mathbf{Z}_2 \times \{0\} = \mathbf{Z}_2 \times \{2\}$, $(1,2) + \mathbf{Z}_2 \times \{0\} = \mathbf{Z}_2 \times \{2\}$.

Here are the calculations for the last coset written out in full.

$(1,2) + \mathbf{Z}_2 \times \{0\} = (1,2) + \{(0,0), (1,0)\} = \{(1,2), (0,2)\} = \mathbf{Z}_2 \times \{2\}$

**3** The left cosets of $4\mathbf{Z}$ are $4\mathbf{Z}$, $1 + 4\mathbf{Z}$, $2 + 4\mathbf{Z}$ and $3 + 4\mathbf{Z}$.

**4** Let $f : \{\text{left cosets}\} \to \{\text{right cosets}\}$ be defined by $f(xH) = (Hx^{-1})$ for all $x \in G$.

*You need to show that f is well defined. That is, you need to show that if xH and yH are the same coset, they must map to the same image. That is $Hx^{-1} = Hy^{-1}$.*

If $xH = yH$, then, by Theorem 46, part (1), $x^{-1}y \in H$. But you could prove a similar theorem for right cosets, that is, a necessary and sufficient condition for the right cosets $Hx$ and $Hy$ to be equal is $xy^{-1} \in H$. Therefore, using this theorem, a necessary and sufficient condition for the cosets $Hx^{-1}$ and $Hy^{-1}$ to be equal is

$\left(x^{-1}\right)\left(y^{-1}\right)^{-1} \in H$. But $\left(x^{-1}\right)\left(y^{-1}\right)^{-1} = x^{-1}y$, so the conditions are identical. Thus, if $xH = yH$, then $Hx^{-1} = Hy^{-1}$. So $f$ is well defined.

*Injection*. If $f(xH) = f(yH)$ then $Hx^{-1} = Hy^{-1}$. Therefore $\left(x^{-1}\right)\left(y^{-1}\right)^{-1} \in H$ as above, so $x^{-1}y \in H$. But then, by Theorem 46 part (1), it follows that $xH = yH$. Therefore $f$ is an injection.

*Surjection*. Let $Hx$ be a right coset of $H$. Then $f\left(x^{-1}H\right)$ $= H\left(x^{-1}\right)^{-1} = Hx$, so $f$ is surjective. Therefore $f$ is a bijection.

**5**   The cosets are of the form $x + \mathbf{Z}$, where $0 \le x < 1$.

**6**   The elements $(x, y)$ of the coset containing the fixed point $(h, k) \in \mathbf{R} \times \mathbf{R}$ have the form $(x, y) = (h, k) + (\lambda a, \lambda b)$ where $\lambda \in \mathbf{R}$. This can be written as $(x, y) = (h, k) + \lambda(a, b)$ which is the parametric form of the straight line through $(h, k)$ in the direction $(a, b)$.

**7**   Suppose that $x \in yH$. Then $x = yh$ for some $h \in H$, and so $y^{-1}x = h \in H$. Now apply the 'if' part of Theorem 46, part (1), where the roles of $x$ and $y$ are reversed. So $yH = xH$, and the result is proved.

**8**   2, 6, 7 and 11 are generators.

**9**   (1) T. See definition of coset. (2) F. See Example 14.6.2. (3) F. See Example 14.6.2. (4) T. Theorem 47.

**10**   Let $a \in H \cap K$, and let the orders of $H$ and $K$ be $h$ and $k$ respectively. By Theorem 49, the order of $a$ divides $h$ and the order of $a$ divides $k$. But $h$ and $k$ are relatively prime, so the order of $a$ divides 1. Therefore $a = e$, so $H \cap K = \{e\}$.

**11**   Since $H \cap K$ is a subgroup of $H$ and a subgroup of $K$, the order of $H \cap K$ divides 56 and 63, by Theorem 48. Therefore the order of $H \cap K$ is 1 or 7. In the first case, $H \cap K = \{e\}$ and is cyclic. In the second case, as 7 is prime, by Theorem 50, $H \cap K$ is cyclic.

## CHAPTER 15

**1**   (1) Since 11 is prime, the only group of order 11 is $C_{11}$.

(2) The orders of the elements must divide nine, the order of the group, so, apart from the identity element, these orders must be 3 or 9.

First suppose that there is an element of order 9. Then $G \cong C_9..$

Now suppose that all the elements of $G$ have order 3, but no element of order 9. Call one of these elements $a$, so that $\{e, a, a^2\} \in G$. Suppose that $b \notin \{e, a, a^2\}$. Then the elements $\{e, a, a^2, b, ba, ba^2, b^2, b^2a, b^2a^2\}$ must all be different. For suppose that $b^i a^j = b^k a^l$, where $i, j, k, l \in \{0, 1, 2\}$. Then $a^{l-j} = b^{i-k}$ or $a^p = b^q$ where $p, q \in \{0, 1, 2\}$, and $p \equiv l - j \pmod 3$ and $q \equiv i - k \pmod 3$. But $b$ cannot be equal to any power of $a$, because

$b \notin \{e, a, a^2\}$. Therefore $i = k$, and hence also $l = j$. Therefore the nine elements $\{e, a, a^2, b, ba, ba^2, b^2, b^2a, b^2a^2\}$ are all different.

Now consider the element $ab$. It must be one of the nine elements, and can only be equal to $ba$, $ba^2$, $b^2a$ or $b^2a^2$.

Case 1. $ab = ba$. The group $G$ is abelian and $G \cong C_3 \times C_3$.

Case 2. $ab = ba^2 = ba^{-1}$. Then $(ab)^2 = (ab)(ab) = (ba^{-1})(ab) = b^2$ and $(ab)^3 = (ab)(ab)^2 = (ab)b^2 = ab^3 = a$ contradicting the fact that the order of $ab$ is three.

Case 3. $ab = b^2a = b^{-1}a$. Then $(ab)^2 = (ab)(ab) = (ab)(b^{-1}a) = a^2$ and $(ab)^3 = (ab)^2(ab) = a^2(ab) = a^3b = b$, contradicting the fact that the order of $ab$ is three.

Case 4. $ab = b^2a^2 = b^{-1}a^{-1}$. Then $(ab)^2 = (ab)(ab) = (ab)(b^{-1}a^{-1})$ $= e$, showing that $ab$ has order two, contradicting the fact that the order of $ab$ is three.

Therefore the only groups of order 9 are $C_9$ and $C_3 \times C_3$.

(3) The orders of the elements must divide the order of the group, 10, so, apart from the identity element, these orders must be 2, 5 or 10.

First suppose that there is an element of order 10. Then $G \cong C_{10}$.

Now suppose that $G$ has no element of order 10. Then there must be an element of order 5, for, if there were not, all the non-identity elements would be of order 2, and, by Theorem 38, the order of $G$ would be a power of 2. So let $a$ be an element of order 5.

If $b \notin \{e, a, a^2, a^3, a^4\}$, then $\{e, a, a^2, a^3, a^4, b, ba, ba^2, ba^3, ba^4\}$ are all distinct elements of $G$. But $b^2 \neq ba, ba^2, ba^3$ or $ba^4$, because $b$ is not a power of $a$. Nor can $b^2$ be $a$, $a^2$, $a^3$ or $a^4$, because that would imply that $b$ is of order 10. This leaves only $b^2 = e$.

The strategy now is to look at the possible outcomes of the product $ab$. Clearly, $ab \neq e, a, a^2, a^3, a^4$ or $b$. So there are now four sub-cases to consider: $ab = ba$, $ab = ba^2$, $ab = ba^3$ and $ab = ba^4$.

Case 1. $ab = ba$. In this case $G$ is abelian and $G \cong C_2 \times C_5$.

Case 2. $ab = ba^2$. Then $(ab)^2 = (ab)(ab) = (ab)(ba^2) = a^3$. Hence $(ab)^3 = (ab)(ab)^2 = ba^2a^3 = b$, $(ab)^4 = (ab)^2(ab)^2 = a^6 = a$ and $(ab)^5 = (ab)^3(ab)^2 = ba^3$. This contradicts the fact that the order of $ab$ is two or five.

Case 3. $ab = ba^3$. In this case $(ab)^2 = (ab)(ab) = (ab)(ba^3) = a^4$; $(ab)^3 = (ab)^2(ab) = a^4(ab) = b$; $(ab)^4 = (ab)^2(ab)^2 = a^8 = a^3$ and $(ab)^5 = (ab)^3(ab)^2 = ba^4$. This contradicts the fact that the order of $ab$ is two or five.

Case 4. $ab = ba^4$. Then $a^5 = b^2 = e$ and $ab = ba^4$, so $G \cong D_5$.

Therefore the only groups of order 10 are $C_{10}$, $C_2 \times C_5$ and $D_5$.

# CHAPTER 16

**1** (1) $(a,b) \sim (a,b)$, since $ab - ab = 0$, so $\sim$ is reflexive. If $(a,b) \sim (c,d)$ then $ad - bc = 0$; so $bc - ad = 0$, and $(c,d) \sim (a,b)$. Therefore $\sim$ is symmetric. Finally, if $(a,b) \sim (c,d)$ and $(c,d) \sim (e,f)$, then $ad - bc = 0$ and $cf - de = 0$, so $adf - bde = d(af - be) = 0$. Suppose that $d \neq 0$. Then $af - be = 0$. Now suppose that $d = 0$. Then either ($b = 0$ or $c = 0$), and ($f = 0$ or $c = 0$). But $c \neq 0$, since $(0,0)$ is not in the original set. Therefore $b = 0$ and $f = 0$, and once again $af - be = 0$. Therefore, in both cases $af - be = 0$, so $(a,b) \sim (e,f)$, and $\sim$ is transitive. Therefore $\sim$ is an equivalence relation. If $(x,y)$ lies in the same equivalence class as $(a,b)$, then $ay - bx = 0$. Therefore the line joining the origin to $(x,y)$ is the same line as that joining the origin to $(a,b)$. Therefore the equivalence classes are the lines through the origin, but none of the lines includes the origin.

(2) This is similar to part (1). The equivalence classes are the elements of **Q**.

(3) Note that $x \sim y$ if, and only if, $x \equiv y \pmod 2$. Then $x \sim x$, since $x \equiv x \pmod 2$, so $\sim$ is reflexive. If $x \sim y$ then $x \equiv y \pmod 2$, so $y \equiv x \pmod 2$, so $y \sim x$ and $\sim$ is symmetric. Finally, if $x \sim y$ and $y \sim z$, then $x \equiv y \pmod 2$ and $y \equiv z \pmod 2$. Therefore $x \equiv z \pmod 2$, so $\sim$ is transitive. The equivalence classes are the even and odd integers.

(4) Note that $x \sim y$ if, and only if, $2x + y \equiv 0 \pmod 3$, and that $2x + y \equiv 0 \pmod 3$ if, and only if, $x \equiv y \pmod 3$. Proving that $\sim$ is an equivalence relation is then similar to part (3), with 2 replacing 3 throughout. The equivalence classes are the integers modulo 3.

(5) $|x| = |x|$, so $\sim$ is reflexive. If $x \sim y$ then $|x| = |y|$, so $|y| = |x|$, and $y \sim x$, so $\sim$ is symmetric. Finally, if $x \sim y$ and $y \sim z$, then $|x| = |y|$ and $|y| = |z|$, so $|x| = |z|$, $x \sim z$ and $\sim$ is transitive. The equivalence classes are the sets $\{\pm n : n \in \mathbf{Z}\}$.

**2** (1) Not an equivalence relation, since $1 \sim 0$ but $0 \nsim 1$, since $-1$ is not a perfect square.

(2) Not an equivalence relation, since $0 \nsim 0$, as $0 \times 0 \ngtr 0$.

(3) This is an equivalence relation, as, trivially, all the conditions are satisfied. There is just one equivalence class, that of $\mathbf{Z}^+$ itself.

(4) Not an equivalence relation, since $1 \sim 0$ and $0 \sim -1$, but $1 \nsim -1$.

(5) Not an equivalence relation, since $0 \sim \frac{1}{2}$ and $\frac{1}{2} \sim 1$, but $0 \nsim 1$.

(6) Not an equivalence relation, since $1 \sim 2$ and $2 \sim 3$, but $1 \nsim 3$.

(7) This is an equivalence relation provided you agree that a line is parallel to itself, so that $l \sim l$, and $\sim$ is reflexive. If $l \sim m$, then $l$ is parallel to $m$, so $m$ is parallel to $l$, and $m \sim l$, and $\sim$ is symmetric. If

$l \sim m$ and $m \sim n$, then $l$ is parallel to $m$, and $m$ is parallel to $n$, so $l$ is parallel to $n$. So $\sim$ is transitive and $\sim$ is an equivalence relation. The equivalence classes are sets of parallel lines in the plane.

(8) Not an equivalence relation, since $(1,1) \nsim (1,1)$.

(9) This is an equivalence relation if you agree that a triangle is similar to itself, so that $A \sim A$, and $\sim$ is reflexive. If $A \sim B$, then $A$ is similar to $B$, so $B$ is similar to $A$, and $B \sim A$, and $\sim$ is symmetric. If $A \sim B$ and $B \sim C$, then $A$ is similar to $B$, and $B$ is similar to $C$, so $A$ is similar to $C$. So $\sim$ is transitive and therefore $\sim$ is an equivalence relation. The equivalence classes are sets of similar triangles.

(10) This is an equivalence relation. The solution is identical to that of part (9), with the word 'congruent' replacing similar.

(11) Not an equivalence relation, since $l$ is not perpendicular to itself.

# CHAPTER 17

**1** (1) $AB = \{ae, aba, be, bba\} = \{a, b\}$

*Notice that there are only two elements as a and b are repeated.*

(2) $AB = \{e, a, a^2, ba^2, b, ba\}$

(3) $AB = \{a, a^2, e, b, ba, ba^2\}$

(4) $AB = \{b, ba, ba^2\}$

**2** $HA = \{a, ba, a^3, ba^3\}$, $HB = \{ba, ba^3\}$, $AB = \{b, ba^2, a^2, e\}$, $AH = \{a, a^3, ba, ba^3\}$, $BH = \{ba, ba^3\}$, $BA = \{ba^2, a^2, b, e\}$.

**3** $aH = \{a, a^2, e\}$, $Hb = \{b, ba^2, ba\}$.

**4** The group $\mathbf{Z}_6 \times \mathbf{Z}_4$ has 24 elements, and the group generated by $(2,2)$, that is, $\langle(2,2)\rangle = \{(2,2), (4,0), (0,2), (2,0), (4,2), (0,0)\}$ has six elements. Therefore $\langle(2,2)\rangle$ has four cosets, so the quotient group $(\mathbf{Z}_6 \times \mathbf{Z}_4)/\langle(2,2)\rangle$ has four elements. You can write these elements as $\langle(2,2)\rangle$, $(1,0) + \langle(2,2)\rangle$, $(0,1) + \langle(2,2)\rangle$ and $(1,1) + \langle(2,2)\rangle$. Since the order of each of these elements, other than the identity $\langle(2,2)\rangle$, is 2, $(\mathbf{Z}_6 \times \mathbf{Z}_4)/\langle(2,2)\rangle \cong \mathbf{Z}_2 \times \mathbf{Z}_2$.

**5** The group $\mathbf{Z}_6 \times \mathbf{Z}_4$ has 24 elements, and the group generated by $\langle(3,2)\rangle$, that is, $\langle(3,2)\rangle = \{(3,2), (0,0)\}$, has two elements. Therefore $\langle(3,2)\rangle$ has 12 cosets, so the quotient group $(\mathbf{Z}_6 \times \mathbf{Z}_4)/\langle(3,2)\rangle$ has 12 elements. You can write these elements as $\langle(3,2)\rangle$, $(1,0) + \langle(3,2)\rangle$, $(2,0) + \langle(3,2)\rangle$, $(3,0) + \langle(3,2)\rangle$, $(4,0) + \langle(3,2)\rangle$, $(5,0) + \langle(3,2)\rangle$, $(0,1) + \langle(3,2)\rangle$, $(1,1) + \langle(3,2)\rangle$, $(2,1) + \langle(3,2)\rangle$, $(3,1) + \langle(3,2)\rangle$, $(4,1) + \langle(3,2)\rangle$ and $(5,1) + \langle(3,2)\rangle$. This is in effect the group $\mathbf{Z}_6 \times \mathbf{Z}_2$, so $(\mathbf{Z}_6 \times \mathbf{Z}_4)/\langle(3,2)\rangle \cong \mathbf{Z}_6 \times \mathbf{Z}_2$

**6**   Take the element $a \in D_3$ and $b \in H$. Then $a^{-1}ba = a^2ba = ba^2$, and as $ba^2 \notin \{e,b\}$, by Theorem 58, $\{e,b\}$ is not a normal subgroup.

**7**   (1) T. (2) F. See Lagrange's Theorem. (3) T.

**8**   For any $g \in Q_4$, $g^{-1}eg = g^{-1}g = e \in H$. Now consider $g^{-1}a^2g$ where $g = b^ia^j$, for $i = 0,1$ and $j = 0,1,2,3$. We need to prove that this belongs to $H$. As $a^2 = b^2$, $g^{-1}a^2g = g^{-1}b^2g$. But $g^{-1}b^2g = \left(b^ia^j\right)^{-1}b^2\left(b^ia^j\right) = a^{-j}b^{-i}b^2b^ia^j = a^{-j}b^2a^j = a^{-j}a^2a^j = a^2$. Therefore $g^{-1}a^2g = a^2$. But $a^2 \in H$, so $g^{-1}a^2g \in H$.

Since every element of $Q_4/H$ is its own inverse, and the group is of order 4, $Q_4/H \cong Z_2 \times Z_2$.

**9**   *If.* Suppose that every right coset of $H$ in $G$ is also a left coset of $H$ in $G$. Thus, if $g \in G$, then, as $Hg$ is a right coset, it must be a left coset. But which left coset? Since $e \in H$, $eg \in Hg$ so $g \in Hg$. But $g$ belongs to the left coset $gH$. Therefore the left coset must be $gH$, so $gH = Hg$. Therefore, for any $h \in H$, $hg = gh_1$ for some $h_1 \in H$. So $g^{-1}hg = h_1 \in H$. So, by Theorem 59, $H$ is a normal subgroup.

*Only if.* Suppose that $H$ is a normal subgroup of $G$ and let $Hx$ be a right coset of $H$ in $G$ for a given element $x \in G$. Let $y \in Hx$. Then $y = hx$ for some $h \in H$. Consider $x^{-1}y$. Since $y = hx$, $x^{-1}y = x^{-1}hx$, and, as $H$ is normal, $x^{-1}hx \in x^{-1}Hx = H$. Therefore $x^{-1}y \in H$, so $x^{-1}y = h_1$ for some $h_1 \in H$, so $y = xh_1$. Thus $y \in xH$, so $Hx \subseteq xH$.

Second part, showing that $xH \subseteq Hx$. Let $y \in xH$. Then $y = xh$ for some $h \in H$. Consider $yx^{-1}$. As $y = xh$, $yx^{-1} = xhx^{-1} = g^{-1}hg$ where $g = x^{-1}$. But, since $H$ is normal, $g^{-1}hg \in H$, so $yx^{-1} \in H$. Therefore $yx^{-1} = h_1$ for some $h_1 \in H$, so $y = h_1x$. Thus $y \in Hx$, so $Hx \subseteq xH$.

Therefore $Hx \subseteq xH$ and $Hx \subseteq xH$, so $Hx = xH$. Therefore every right coset of $H$ in $G$ is also a left coset of $H$ in $G$.

# CHAPTER 18

**1**   Let $A$ be a subgroup of $G$, and let the image of $A$ under $f$ be $B$. Suppose that $x, y \in B$. Then, since $B$ is the image of $A$, there exist $a, b \in A$ such that $f(a) = x$ and $f(b) = y$. Then $xy = f(a)f(b)$, and since $f$ is a homomorphism, $f(a)f(b) = f(ab)$. Hence $xy = f(ab)$ and is therefore in the image of $A$ under $f$, that is, $B$.

By Theorem 61 $f(e_G) = e_H$. But, by Theorem 21, $e_G \in A$, so the image of the identity, $e_H \in B$.

Suppose that $x \in B$. Let $x$ be the image of $a \in A$ so that $f(a) = x$. Then, as $A$ is a subgroup of $G$, $a^{-1} \in A$. Then $e_H = f(e_G) = f(a^{-1}a) = f(a^{-1})f(a) = f(a^{-1})x$, so $e_H = f(a^{-1})x$. Similarly, $e_H = xf(a^{-1})$. Therefore $f(a^{-1})$ is the inverse of $x$, and since $a^{-1} \in A$, $f(a^{-1}) \in B$.

Therefore, by Theorem 21, $B$ is a subgroup of $H$.

**2**  Let $g$ be a generator of $G$, and let $f(g) = h$. You will find a proof that $f(g^r) = h^r$ for all $r \in \mathbf{Z}$ in the answer to question 13 in Chapter 11. (The argument there did not use the fact that the homomorphism $f$ was actually an isomorphism.)

Given $y \in H$, there exists $x \in G$ such that $f(x) = y$, since $f$ is surjective. As $G = \langle g \rangle$, $x = g^n$ for some $n \in \mathbf{Z}$. So $y = f(x) = f(g^n) = h^n$. Therefore $h$ is a generator for $H$, so $H$ is cyclic.

**3**  As $f(m)f(n) = a^m a^n = a^{m+n} = f(m+n)$, $f$ is a homomorphism.

**4**  (1) A homomorphism, since $f(x)f(y) = |x||y| = |xy| = f(xy)$. The kernel consists of $x$ such that $|x| = 1$, that is, $\{1, -1\}$.

(2) Not a homomorphism, since $f(3.7) + f(3.7) = 3 + 3 = 6$, and $f(3.7 + 3.7) = f(7.4) = 7$.

(3) Not a homomorphism, since $f(1) + f(1) = (1+1) + (1+1) = 4$ and $f(1+1) = 2 + 1 = 3$.

(4) A homomorphism, since $f(x)f(y) = (1/x)(1/y) = (1/xy) = f(xy)$. The kernel consists of $x$ such that $f(x) = 1/x = 1$, that is, $\{1\}$.

**5**  Let $x, y \in G$. Then $(Ff)(x)(Ff)(y) = F(f(x))F(f(y)) = F(f(x)f(y)) = F(f(xy)) = (Ff)(xy)$, so $(Ff)$ is a homomorphism.

**6**  *If.* Suppose that $\ker f = \{e_G\}$. *Injection.* Suppose that $f(x) = f(y)$. Then, by Theorem 63, $x$ and $y$ belong to the same coset of $\ker f$ in $G$. But $x \ker f = \{x\}$, and if $y \in x \ker f$, $y = x$. By hypothesis, $f$ is an injection, and therefore $f$ is an isomorphism.

*Only if.* Suppose that $f$ is an isomorphism. Then, by Theorem 61, $f(e_G) = e_H$, so $e_G \in \ker f$. But $f$ is also a bijection, so no other element maps to $e_H$. Therefore $\ker f = \{e_G\}$.

## CHAPTER 19

**1**  Define $f : S_n \to (\{1, -1\}, \times)$ by $f(x) = (-1)^{N(x)}$. This function is surjective, as $f((12)) = -1$, so $\operatorname{im} f = \{1, -1\}$.

*Homomorphism.* Using Theorem 41, $f(x)f(y) = (-1)^{N(x)}(-1)^{N(y)} = (-1)^{N(x)+N(y)} = (-1)^{N(xy)} = f(xy)$, so $f$ is a homomorphism.

Then, since $f$ is a homomorphism and $A_n$ is its kernel, $A_n$ is a normal subgroup of $S_n$, and $S_n / A_n \cong (\{1, -1\}, \times)$.

**2**  Define $f : (\mathbf{C}^*, \times) \to (\mathbf{R}^+, \times)$ by $f(z) = |z|$. This function is surjective, because, if $x \in \mathbf{R}^+$, $f(x + 0i) \to x$, so $\operatorname{im} f = \mathbf{R}^+$.

*Homomorphism.* $f(z_1)f(z_2) = |z_1||z_2| = |z_1 z_2| = f(z_1 z_2)$.

$\ker f = T$, where $T = \{z \in \mathbf{C}^* : |z| = 1\}$, the unit circle subgroup of $\mathbf{C}^*$. Thus, $(\mathbf{C}^*, \times)/T \cong (\mathbf{R}^+, \times)$.

**3**    The image of $f$ is $\mathbf{R}$, for if $k \in \mathbf{R}$, then $f(k\mathbf{a}/\mathbf{a}.\mathbf{a}) = (k\mathbf{a}/\mathbf{a}.\mathbf{a}).\mathbf{a}$ $= k\mathbf{a}.\mathbf{a}/\mathbf{a}.\mathbf{a} = k$.

*Homomorphism.* $f(\mathbf{x}) + f(\mathbf{y}) = \mathbf{a}.\mathbf{x} + \mathbf{a}.\mathbf{y} = \mathbf{a}.(\mathbf{x} + \mathbf{y}) = f(\mathbf{x} + \mathbf{y})$

The kernel is the plane in $\mathbf{R}^3$ through the origin perpendicular to $\mathbf{a}$; call this plane $\Pi_{\mathbf{a}}$. Therefore $\mathbf{R}^3/\Pi_{\mathbf{a}} \cong \mathbf{R}$.

**4**    Define $f : G \to \mathbf{Z}_2 \times \mathbf{Z}_2$ by $f(a + bi) = ([a]_2, [b]_2)$.

The image of $f$ is $\mathbf{Z}_2 \times \mathbf{Z}_2$, since, for any $([a]_2, [b]_2) \in \mathbf{Z}_2 \times \mathbf{Z}_2$, $f(a + bi) = ([a]_2, [b]_2)$.

*Homomorphism.*

$$f(a + bi) + f(c + di) = ([a]_2, [b]_2) + ([c]_2, [d]_2)$$

$$= ([a]_2 + [c]_2, [b]_2 + [d]_2)$$

$$= ([a + c]_2, [b + d]_2)$$

$$= f((a + c) + (b + d)i)$$

$\ker f = \{a + bi : f(a + bi) = ([0]_2, [0]_2)\}$, that is $a + bi \in \ker f$ if, and only if, $a, b \in 2\mathbf{Z}$. Let $H = \{a + bi : a, b \in 2\mathbf{Z}\}$.

Then, by the first isomorphism theorem, $G/H \cong \mathbf{Z}_2 \times \mathbf{Z}_2$.

**5**    Define $f : \mathbf{Z}_{pq} \to \mathbf{Z}_p$ by $f([a]_{pq}) = [a]_p$.

*Well defined.* If $[a]_{pq} = [b]_{pq}$, then $pq$ divides $a - b$. Therefore $p$ divides $a - b$, so $[a]_p = [b]_p$. Therefore $f$ is well defined.

The image of $f$ is $\mathbf{Z}_p$, because given $[a]_p \in \mathbf{Z}_p$, $f([a]_{pq}) = [a]_p$.

*Homomorphism.*

$$f([a]_{pq}) + f([b]_{pq}) = [a]_p + [b]_p$$

$$= [a + b]_p$$

$$= f([a + b]_{pq})$$

The kernel is the set of elements $[a]_{pq}$ such that $f([a]_{pq}) = [0]_p$, that is the set of elements $[a]_{pq} \in \mathbf{Z}_{pq}$ such that $a$ is divisible by $p$. Hence $\ker f = \{[0]_{pq}, [p]_{pq}, [2p]_{pq}, \ldots, [(q-1)p]_{pq}\}$, so $\ker f = (p\mathbf{Z})_{pq}$. Therefore, $\mathbf{Z}_{pq}/(p\mathbf{Z})_{pq} \cong \mathbf{Z}_p$.

# Index